WIRELESS COMMUNICATION USING DUAL ANTENNA ARRAYS

THE KLUWER INTERNATIONAL SERIES
IN ENGINEERING AND COMPUTER SCIENCE

WIRELESS COMMUNICATION USING DUAL ANTENNA ARRAYS

by

Da-shan Shiu
QUALCOMM, INC., U.S.A.

KLUWER ACADEMIC PUBLISHERS
Boston / Dordrecht / London

Distributors for North, Central and South America:
Kluwer Academic Publishers
101 Philip Drive
Assinippi Park
Norwell, Massachusetts 02061 USA
Telephone (781) 871-6600
Fax (781) 871-6528
E-Mail <kluwer@wkap.com>

Distributors for all other countries:
Kluwer Academic Publishers Group
Distribution Centre
Post Office Box 322
3300 AH Dordrecht, THE NETHERLANDS
Telephone 31 78 6392 392
Fax 31 78 6546 474
E-Mail <orderdept@wkap.nl>

 Electronic Services <http://www.wkap.nl>

Library of Congress Cataloging-in-Publication Data

A C.I.P. Catalogue record for this book is available
from the Library of Congress.

99-47345

To my parents

CONTENTS

Foreword

At present, the expansion of tetherless communications is a technological trend surpassed perhaps only by the explosive growth of the Internet. Wireless systems are being deployed today mainly for telephony, satisfying the industrialized nations' appetite for talk-on-the-go, and providing much-needed communications infrastructure in developing countries. The desire for wireless access to the Internet is starting to add fuel to the growth of tetherless communications. Indeed, the synergy of wireless and Internet technologies will lead to a host of exciting new applications, some of which are not yet envisioned.

Future-generation wireless systems will achieve capacities much higher than the systems of today by incorporating myriad improvements. These innovations include transmission in higher-frequency bands, "smart antennas", multi-user detection, new forward error-correction techniques, and advanced network resource-allocation techniques.

The term "smart antenna" usually refers to the deployment of multiple antennas at the base-station site, coupled with special processing of the multiple received signals. Smart antennas can adaptively reject co-channel interference and mitigate multipath fading, and have been identified by many as a promising means to extend base-station coverage, increase system capacity and enhance quality of service.

Currently, smart antennas are added to existing systems in a way that is transparent to the rest of the network from an operational perspective. This approach is cost-effective and backward-compatible. Over the next few years, much of the new wireless infrastructure will employ smart antennas, and to fully utilize their potential, the air-interface standards must be designed with smart antennas in mind. For example, proposed third-generation cellular radio standards include built-in support for smart antennas.

Looking further ahead, once smart antennas are widely deployed and their benefits have been exploited fully, what should we do next? In this book, Dashan Shiu helps answer this question by exploring the use of antenna arrays at both ends of the wireless link. Dual-antenna array systems offer unprecedented spectral efficiencies over wireless channels in which a line-of-sight path is not present. Since the announcement by Lucent Technologies of their

BLAST (Bell-labs LAyered Space-Time) prototype, dual-antenna array systems have garnered considerable attention from both industrial and academic researchers.

Dual-antenna arrays will become a key technology for future wireless systems, providing enormous capacity increases that will enable high-speed mobile Internet access, enhanced-capacity wireless local loops, wireless high-definition video transport, and other exciting applications. Unlike telephony applications, which require a constant bit rate per user, many of these new applications generate bursty, asynchronous traffic, which is well-matched to the high average throughput and very high peak throughput provided by dual-antenna arrays.

The theory of dual-antenna array systems is not a straightforward extension of the existing theory of single antenna-to-multiple antenna communications, and a single volume cannot hope to provide encyclopedic coverage of dual-antenna array theory. This book treats several key topics in depth, including signal propagation, transmit power allocation, information-theoretic channel capacity, and coding and decoding techniques. Da-shan Shiu discusses the "what-is", "what-to-expect" and "how-to" of dual-antenna array systems. An important unifying theme is how to exploit, rather than mitigate, multipath fading effects. While this might seem counterintuitive at first, Chapter 3 should persuade even the skeptical reader of the beneficial effects of multipath fading.

I recommend this book highly.

Joseph M. Kahn
Berkeley, California
August 10, 1999

Preface

Dual antenna-array systems offer a significantly larger channel capacity than single-antenna systems. As a rule of thumb, if the fades between pairs of transmit-receive antennas are i.i.d., the average channel capacity of a dual antenna-array system that uses n antennas at both the transmitter and the receiver is approximately n times higher than that of a single-antenna system. Furthermore, this increased spectral efficiency cannot be obtained by any other known methods. This book investigates a few fundamental issues with wireless communications using dual antenna arrays.

The material in this book is primarily intended for engineers, scientists, and so forth who want to start learning about this exciting new paradigm. We assume a basic knowledge of signal processing, linear system theory, digital communications, and information theory.

A GUIDED TOUR OF THE BOOK

Mathematical Preliminaries

Chapter 2 establishes a generic mathematical representation for a multiple-input, multiple-output (MIMO) frequency-nonselective Rayleigh fading channel. The expression for channel capacity assuming that the receiver has a perfect measurement of the channel is presented.

We also provide the asymptotic property of channel capacity assuming that the channel fades are independent. It is clear that the channel capacity scales linearly with the number of antenna elements.

Fading Correlation Model

A fading channel is considered as a random variable. Its distribution not only determines the distribution of channel capacity but also the overall signal processing architecture. In Chapter 3 we present an abstract model for the multipath propagation environment of a typical outdoor fixed wireless link and derive the corresponding channel distribution, or fading correlation. This model is a reasonable approximation of the real propagation environment and is simple enough to lead to useful insights, such as the relationship between effective degrees of freedom and fading correlation, about MIMO fading channels. We then discuss the dependence of channel capacity on model parameters such as angle spread, antenna element spacing, and angle of arrival.

The MIMO channel can be decomposed into a set of equivalent single-input, single-output subchannels. The effect of severe fading correlation is to reduce the number of subchannels that are active in conveying information.

Power Allocation Strategies

A power allocation strategy determines the allocation of physical transmit power, and hence the communication rate, to each spatial dimension of the transmitted signal. Choosing an appropriate power allocation strategy is particularly important when the SNR is low.

In Chapter 4, we discuss three power allocation strategies for dual antenna-array systems, assuming that the channel can be described by the model in Chapter 3. Optimum power allocation achieves the highest capacity but requires the transmitter to have instantaneous channel state information (CSI). When the transmitter does not have CSI, uniform power allocation can be employed. It performs well when the fading correlation is low. When the fading correlation is high, stochastic water-filling power allocation performs very close to optimum power allocation in the downlink, and uniform power allocation achieves the highest average capacity in the uplink.

Layered Space-Time Codes

Space-time codes are channel codes with multiple spatial dimensions that can be used to utilize the high channel capacity of dual antenna-array systems. In particular, such codes are necessary in systems in which the transmitter does not have the instantaneous CSI. Unfortunately, the decoding complexity of space-time codes can be very high. We propose layered space-time (LST) codes which allow for low-complexity decoding. We analyze the performance of LST codes and define the key design parameters to formulate the design criteria for LST codes. Furthermore, we examine the use of convolutional codes and block codes as the constituent codes for DLST codes, and propose modifications to the LST architecture to greatly improve the performance of LST codes.

Transmit Diversity

In many applications, antenna arrays are deployed only at the base stations due to some physical and cost considerations. In Chapter 6 we discuss techniques that can be applied to improve the quality of transmission from an antenna array to a single antenna. We provide a channel capacity analysis to evaluate the performance of these transmit diversity schemes.

Open Issues

The final chapter is a personal statement about the future research directions on dual antenna-array systems. We identify a few key areas that warrant further research and development: further understanding of channel statistics, acquisition and tracking of CSI, signal processing techniques, network issues, distributed BS antenna scheme, and high performance space-time codes.

ACKNOWLEDGMENTS

I am deeply indebted to the many people who have helped me immeasurably on this work. In particular I owe a debt of gratitude to Prof. Joseph Kahn. Special thanks to Drs. Jack Salz, Gerard Foschini, Mike Gans, and Reinaldo Valenzuela for introducing me to the very topic of wireless communications using dual antenna arrays. I also thank Prof. Venkat Anantharam and Prof. David Tse for their invaluable advice.

I would like to acknowledge and thank the following individuals: the Chao family, Dr. Tao Chen, Dr. Mui-Choo Chuah and Chen-Nee Chuah, Dr. Gene Marsh, Prof. Ricky Ho, Bilung Lee, Dr. David Lee, Jocelyn Nee, Dr. Tomoaki Ohtsuki, and Pramod Viswanath.

Two Funds from the National Science Foundation and the University of California MICRO Program supported my research at UC Berkeley.

1

Introduction

1.1 Dual Antenna Array Systems

The demand for higher data rates and higher quality in wireless communication systems has recently seen unprecedented growth. One of the most limiting factors in wireless communications is the scarcity of spectrum. Techniques that improve spectral efficiency, such as the cellular structure that allows frequency reuse, have had tremendous impact on the proliferation of wireless communications. In this book, we explore a new technology that can dramatically increase the spectral efficiency. One key element of this technology is to use antenna arrays at both the transmitter and receiver.

Antenna arrays have been used to combat various types of channel impairments. An antenna array with sufficient antenna spacing can provide spatial diversity to mitigate multipath fading. Beamforming and diversity reception can be employed to combat the effect of delay spread and co-channel interference. For a comprehensive summary, see [1]. The advances in technology and the expanding demands for antenna arrays have made them very economical, and the trend of using GHz carriers for wireless access networks reduce the size requirements of the antenna arrays.

One implicit assumption underlying these traditional uses of antenna arrays is that the information content transmitted or received by each antenna element is identical. This is intuitive because if unrelated signals are transmitted from different antenna elements, these signal will interfere with each other at the receiver. As an example, consider beamforming. By properly adjusting the phase of each antenna, the main lobe of the antenna pattern can be directed to the desired angle. This enhances the strength of the desired signal and also suppresses the interference from signals coming from undesired angles. However, we will show in this book that the key to fulfilling the potential of dual antenna array systems is to transmit independent information from each antenna element.

It is also believed that maintaining a direct line-of-sight path between the transmitter and the receiver is desirable because it minimizes the scattering and absorption of the signal. In this book we will also show that in certain situations the multipath fading introduced by the scatterers can indeed lead to a channel capacity much higher than if the channel is line-of-sight.

1.2 Systems Having Multiple Antennas at Both the Transmitter and the Receiver

Recently, a radically different paradigm for the use of antenna arrays has been proposed. It has been shown that wireless systems using multiple antennas at both the transmitter and the receiver offer a large capacity. As a rule of thumb, if the fades between pairs of transmit-receive antennas are i.i.d., the average channel capacity of a dual antenna-array system that uses n antennas at both the transmitter and the receiver is approximately n times higher than that of a single-antenna system for a fixed bandwidth and overall transmitted power [2] - [4]. Furthermore, this increased spectral efficiency cannot be obtained by any other known methods.

The following are the requirements for such a high channel capacity.

- Antenna arrays with sufficient spacings must be deployed at both ends.

- The link must employ no conventional mechanism, such as frequency- or code- division multiplexing, to ensure that the signals transmitted by different transmitting antennas are orthogonal to each other at the receiver.

- The propagation environment between the transmitter and the receiver must provide numerous propagation paths.

- The receiver must be able to measure or estimate the channel gain, both amplitude gain and phase shift. To our knowledge, to date the proposed detection techniques require the receiver to apply coherent processing techniques over the received signals.

There are a large number of potential applications for such dual antenna array systems [5]. However, because of the need for a good channel estimation and the required size or form factor of the antenna arrays, in our opinion applications using the first generation of implementations will be restricted to low-mobility, medium-sized client terminals.

This finding of the new paradigm has spurred a great interest in the communications research community. Initial results on channel measurement, channel capacity, channel modeling and simulation, space-time signal processing techniques and space-time channel codes, equalization, and prototyping have been reported. We expect to see much more research and development activities on dual antenna-array systems with space-time processing in the near future, because this is truly an innovation with a tremendous impact on wireless communications.

1.3 Overview

The next chapter presents the fundamentals of wireless communications in fading environments. We describe a generic mathematical representation for a multiple-input, multiple-output (MIMO) frequency-nonselective Rayleigh fading channel. We then present formulas for calculating channel capacity subject to constraints on the second-order statistics of the input signal assuming that the receiver has a perfect measurement of the channel. To serve as a motivating point, we provide a proof of the almost-sure convergence of per-antenna capacity, which is equivalent to saying that the expected value of channel capacity grows linearly in the number of antenna, under an idealized assumption on the spatial fading correlation. We will also present a result which demonstrates the importance of having the channel measurement at the receiver. If the channel measurement is unavailable to the receiver, the linear growth of channel capacity with respect to the number of antenna elements ceases after the number of antenna elements reach the coherence time of the channel.

The design of dual antenna-array systems and the analysis of their performance require a new class of channel models that pay more attention to the spatial characteristics. In Chapter 3, we will present a scatterer model which is appropriate in the context where one end of the wireless link is elevated and unobstructed while the other end is surrounded by local scatterers. From this model, we show how the spatial fading correlation can be reasonably estimated given key physical parameters such as angle spread and angle of arrival. In explaining the effect of spatial fading correlation on the channel capacity of dual antenna-array systems, we decompose the MIMO channel into an equivalent system consisting of a set of parallel single-input, single-output (SISO) channels and show that the fading correlation modifies the gain distributions of these SISO channels.

Chapter 4 discusses the performance of dual antenna-array systems with different power allocation strategies. The term "power allocation strategy" refers to how the transmitted power is distributed among the n spatially orthogonal transmit modes. Although the capacity using optimum power allocation is the highest, instantaneous channel state information (CSI) at the transmitter is required to implement optimum power allocation. Uniform power allocation, on the other hand, is robust and amenable to implementation. Using the model developed in Chapter 3, we evaluate the performance of these two power allocation strategies. When the fading correlation is high, there is a significant difference between the capacities achievable by the two power allocation strategies. A novel power allocation strategy, which we refer to as stochastic water-filling, is proposed. We show that in the downlink direction the power allocation computed using the stochastic water-filling algorithm yields a capacity significantly higher than uniform power allocation.

Though the high spectral efficiency promised by dual antenna array systems is very exciting, ML detection at the receiver generally requires a complexity that increases exponentially in n. In Chapter 5 we consider a class of channel codes, the layered space-time (LST) codes, whose encoding/decoding complexity increases much less rapidly in n. We first review the LST architecture. Subsequently, we analyze the error performance of LST codes. Based on this analysis, we identify the key parameters of LST codes, and use them to formulate a set of design criteria for LST codes. Next, we present the optimal trade-off among design parameters, and the power penalty associated with suboptimal LST decoding compared to ML decoding. Operational details of employing block and convolutional codes as the constituent codes are examined. We propose modifications to the original LST architecture to greatly enhance the performance of LST codes.

In many existing applications, e.g. cellular radio networks, multiple antenna elements can only be deployed at one end; a single antenna element is used at the other end. While techniques to utilize the receive diversity provided by the antenna array is well documented, effective ways to utilize the transmit diversity, in particular in non-line-of-sight environments, have not been studied until recently. In Chapter 6 we first examine the potential of transmit diversity when there is only one receiving antenna element. We then describe methods that achieve the full or partial benefit of transmit diversity in situations where the transmitter has no channel state information at all.

Chapter 7 summarizes the conclusions of this book and present topics for future research.

2

Background

In this chapter, we discuss the fundamental aspects of wireless communications using multiple antennas at both ends of a link in fading environments. We describe a mathematical representation for a multiple-input, multiple-output (MIMO) frequency-nonselective Rayleigh fading channel. We then examine the definition of channel capacity on such channels assuming that the channel is quasi-static, and provide a proof of the almost sure convergence of channel capacity per antenna. It will become clear that multipath fading is the key to such a high channel capacity. While having the channel state information (CSI) at the receiver is critical, having CSI at the transmitter is not. Furthermore, orthogonalization of signals that carry independent data streams in time, frequency, or code is unnecessary if the number of data streams is less than the number of antenna elements.

Before we proceed, note that the following notation is used throughout this book.

- We focus on single-user to single-user communication using antenna arrays at both ends. We use n and m to denote the number of antennas at the transmitter and the receiver, respectively, and refer to an n-transmit, m-receive antenna system as an (n, m) system.

- Bold-face lower-case letters, e.g. x, refers to column vectors. Upper-case letters, e.g. X, refer to matrices.

- For a matrix X, its lth column is denoted by x_l. That is, $X = \begin{bmatrix} x_0 & x_1 & \ldots \end{bmatrix}$. The kth row, lth column element is denoted by x_l^k or $X(k, l)$.

- We use $*$ for complex conjugate transpose, $'$ for transpose, and $\dagger$ for conjugate transpose.

- A real Gaussian random variable with mean ζ and variance σ^2 is denoted as $N(\zeta, \sigma^2)$. A circularly symmetric complex Gaussian random variable z, denoted by $z \sim \tilde{N}(0, \sigma^2)$, is a random variable $z = x + iy$ in which x and y are i.i.d. $N(0, \sigma^2/2)$.

2.1 Channel and Noise Model

In this book, we assume that the communication is carried out using bursts (packets), and that the channel varies at a rate slow enough that it can be regarded as essentially fixed during a burst. Under this assumption, the multiple-input, multiple-output (MIMO) channel can be regarded as linear time-invariant during a burst transmission. Denote the signal transmitted by the lth transmit antenna by $s^l(t)$ and the signal received by the kth receive antenna by $r^k(t)$. Also denote the impulse response connecting the input of the channel from transmit antenna l to the output of the channel to receive antenna k by $h_l^k(t)$. The input/output relation of the MEA system is described by the following vector notation:

$$r(t) = H(t)*s(t) + v(t), \tag{2-1}$$

where $r(t) = (r^1(t) \ r^2(t) \ \ldots \ r^n(t))'$, $s(t) = (s^1(t) \ s^2(t) \ \ldots \ s^m(t))'$, $H_l^k(t) = h_l^k(t)$, $v(t)$ is additive white Gaussian noise (AWGN), and $*$ denotes convolution.

If the communication bandwidth is narrow enough that the channel frequency response can be treated as flat across frequency, the gain connecting transmitting antenna l and receiving antenna k can be denoted by a complex number h_l^k. The discrete-time system corresponding to (2-1) is:

$$r_\tau = Hs_\tau + v_\tau, \tag{2-2}$$

where τ is the discrete-time index, $s_\tau = (s_\tau^1 \; s_\tau^2 \; ... s_\tau^n)'$, $r_\tau = (r_\tau^1 \; r_\tau^2 \; ... r_\tau^m)'$, $v_\tau = (v_\tau^1 \; v_\tau^2 \; ... v_\tau^m)'$, and the channel matrix is $H(k, l) = h_l^k$.

In this book, we consider Rayleigh fading channels. For Rayleigh fading channels, the channel gain h_l^k is modeled as $\tilde{N}(0, 1)$ [6]. The m noise components of v_τ are assumed to be i.i.d. $\tilde{N}(0, \sigma_v^2)$. The average signal-to-noise ratio (SNR) is defined as $E[s_\tau^\dagger s_\tau]/\sigma_v^2$. To facilitate a fair comparison between systems having different number of antennas, the average SNR, $tr(E[s_\tau^\dagger s_\tau])/\sigma_v^2$, where tr denotes trace, is limited to be no greater than ρ, regardless of n.

In most of this book, we focus on slowly varying, flat Rayleigh fading channels because even with this simple and manageable channel model most important insights into dual antenna-array systems can be obtained. This assumption is relaxed in Chapter 5, where we consider the performance of space-time codes over fast-fading channels. The results obtained here for slowly varying, flat Rayleigh channels can be extended to other classes of channels, using established techniques. For example, a wideband channel with frequency selective fading can be transformed into parallel narrowband channels through the use of orthogonal frequency division multiplexing (OFDM) [7], and the theories described in this book can be applied to each of the narrowband channel.

2.2 Channel Capacity

In the introduction, we mentioned the assumption that communication is carried out using bursts (packets). The burst duration is assumed to be short enough that the channel can be regarded as essentially fixed during a burst, but long enough that the standard information-theoretic assumption of infinitely long code block lengths is a useful idealization. These assumptions are met in, for instance, fixed wireless and indoor wireless applications. In short, for each burst transmission, the channel is randomly drawn from an underlying distribution and stays fixed for the duration of the entire burst. In this quasi-static scenario, it is meaningful to associate a channel capacity with a given realization of channel matrix H. Because the channel capacity is a function of the channel realization, the channel capacity is a random quantity whose distribution is determined by the distribution of H.

Assuming that receiver has perfect CSI, the channel capacity of a communication system described by (2-2) given the channel realization H subject to the constraint that $E[s_\tau s_\tau^\dagger] = \Sigma_s$ is ([2], [8])

$$C = \log_2 \left[\det \left(I + \frac{1}{\sigma_v^2} H \Sigma_s H^\dagger \right) \right] \text{ (bits per channel use)}, \qquad (2\text{-}3)$$

achieved by zero-mean circularly symmetric complex Gaussian input $\tilde{N}(0, \Sigma_s)$. A more relaxed condition is that only the average overall transmitted power, $tr(\Sigma_s)$, is constrained. Subject to this constraint, the channel capacity is

$$C = \max_{tr(\Sigma_s) \le \rho \sigma_v^2} \log_2 \left[\det \left(I + \frac{1}{\sigma_v^2} H \Sigma_s H^\dagger \right) \right] \text{ (bits per channel use)}, \qquad (2\text{-}4)$$

achieved by zero-mean circularly symmetric complex Gaussian input whose covariance matrix is the argument that maximizes (2-4).

In this book, we will present channel capacities subject to various constraints[1] on the second order statistics of the input. Usually the underlying constraint will be obvious from the context. Whenever there is potential for confusion, we will explicitly indicate the particular constraint.

In later chapters we will compare channel capacities subject to different constraints. To compare one channel capacity distribution from another, note that in slow fading environments an important performance measure for an dual antenna-array system is the capacity at a given outage probability q, denoted by C_q. To be specific, the capacity is less than C_q with probability q. In this book, comparisons among different capacity distributions will be presented, when possible, based on the capacity at ten-percent outage, $C_{0.1}$. How-

---

[1]. For a scalar AWGN channel, when the term channel capacity is used, it usually refers to the maximal mutual information subject to a maximal variance constraint on the input. However, for multi-dimensional inputs, consensus on what is the implied constraint has not been reached. Therefore, we feel that it is appropriate not to reserve the term channel capacity for a certain constraint. Instead, channel capacity is used as a synonym for maximal mutual information.

ever, when it is not practical to compute C_q, we will use the expected value of capacity for comparison purposes.

2.3 Asymptotic Behavior of Channel Capacity

We have mentioned in Chapter 1 that the capacity of an (n, m) wireless link assuming idealized i.i.d. flat Rayleigh fading is approximately proportional to $\min(n, m)$. This large capacity is a major motivation for the research described in this book. In the following, we provide a succinct outline of the theoretical derivation of this result. For more details, the reader is referred to [9].

Specifically, we consider an (n, n) i.i.d. flat Rayleigh fading channel. We focus on the ratio between the expected value of channel capacity, $E[C]$, to the number of antennas, n, subject to a constraint on the second-order statistics of the transmitted signal s_τ, $\Sigma_s / \sigma_v^2 = \rho / n \cdot I_n$. Given such a constraint, channel capacity is achieved by a zero-mean circularly symmetric complex Gaussian input distribution $\tilde{N}(0, \rho \sigma_v^2 / n \cdot I_n)$. The channel capacity is

$$C = \log_2 \left[\det \left(I + \frac{\rho}{n} HH^\dagger \right) \right] = \sum_{k=1}^{n} \log_2 \left(1 + \frac{\rho}{n} \varepsilon_k^2 \right), \qquad (2\text{-}5)$$

where ε_k^2 are the eigenvalues of $HH^\dagger$. It has been shown that, as $n \to \infty$, the distribution of the eigenvalues of $HH^\dagger / n$ converges to the following deterministic function almost surely, where λ denotes the eigenvalue of $HH^\dagger / n$ by [10]:

$$g(\lambda) = \begin{cases} \dfrac{1}{\pi} \sqrt{\dfrac{1}{\lambda} - \dfrac{1}{4}}, & 0 \leq \lambda \leq 4, \\ 0 & \text{otherwise.} \end{cases} \qquad (2\text{-}6)$$

Therefore, almost surely,

$$\frac{C}{n} \to \int_0^4 \log_2(1 + \rho\lambda) g(\lambda) d\lambda \equiv C_{\text{uni}}^*, \qquad (2\text{-}7)$$

i.e., C/n converges to a constant C_{uni}^* determined by the SNR ρ.

Note that this convergence result does not require the transmitter to have channel state information because in this case the covariance matrix of the input signal is channel independent.

2.4 Channel capacity when the receiver does not have CSI

Here we examine the channel capacity when neither the transmitter nor the receiver knows the CSI. It says that, if the receiver does not have CSI, increasing the number of transmit antennas beyond the channel coherence time does not improve channel capacity assuming that the channel fades are i.i.d. Rayleigh. The result is due to Marzetta and Hochwald [11].

The channel is assumed to remain constant for T symbol periods $\tau = 1, 2, ..., T$, after which they change to new independent random values which they maintain for another T symbol periods, and so on. T is referred to as the coherence time of the channel. In the following, we assume that m is always larger than n.

Define $\quad S = \left[s_1\ s_2\ \ldots\ s_T \right]'$, $\qquad R = \left[r_1\ r_2\ \ldots\ r_T \right]'$, $\qquad$ and

$V = \left[v_1\ v_2\ \ldots\ v_T \right]'$. Then

$$R = HS + V. \tag{2-8}$$

Because the receiver does not have CSI, the receiver can no longer calculate the likelihood of S given H. Instead, it must formulate the unconditional likelihood of the transmitted signal, which is the conditional probability density of R given S. If the entries of H are i.i.d. $\tilde{N}(0, 1)$,

$$p(R|S) = \frac{\exp(-\text{tr}\{[I_T + (\rho/n)SS^\dagger]^{-1}RR^\dagger\})}{\pi^{Tm}\det^m[I_T + (\rho/n)SS^\dagger]}, \tag{2-9}$$

where I_T denotes the $T \times T$ identity matrix and "tr" denotes trace. Note that the likelihood of a transmitted signal S depends only on $SS^\dagger$. Equation (2-9) suggests that it is the probability distribution of $SS^\dagger$ which is important, not the distribution of S itself.

Consider a particular codebook CB_1. The codebook leads to a probability mass function of $SS^\dagger$. Now construct another codebook CB_2. For each codeword S belonging to CB_1, we put a corresponding codeword L, which is a $T \times T$ lower triangular matrix, in CB_2. These two matrices S and L are related by the Cholesky factorization, $SS^\dagger = LL^\dagger$. Because CB_2 contain only $T \times T$ lower triangular matrices, an antenna array of T elements is enough to transmit the codewords of CB_2. Furthermore, it is shown in [11] that CB_1 and CB_2 generate the same mutual information and conform to the same transmit power constraint:

$$\frac{1}{T}\sum_{\tau=1}^{T}\sum_{k=1}^{n} E\left|s_\tau^k\right|^2 = \rho. \tag{2-10}$$

The conclusion is that increasing the number of transmit antenna does not increase capacity. It is in sharp contrast with the case when the receiver has CSI.

2.5 Discussion

Here we examine the requirements listed in Chapter 1 for the high channel capacity offered by employing dual antenna arrays.

1. Multiple antennas must be deployed at both ends. To be more specific, if $n = o(m)$ (or $m = o(n)$), i.e. the number of antenna elements at one end of a link is insignificant compared to that at the other end, the channel capacity can not grow linearly with m (or n). The proof, assuming that the entries of H are i.i.d., is shown in the Appendix.

2. The propagation environment between the transmitter and the receiver exhibits rich multipath.
   It has been shown in [7] that if the channel consists of only L multipath component from the transmitting antenna array to the receiving antenna array, and that if the distance between the antenna arrays and the scatterers are sufficiently larger than the physical dimension of the antenna arrays, once $\min(m, n)$ is greater than L, the capacity stops increasing linearly with the number of antennas for a reasonable SNR. In this book,

we always assume that the propagation environment provides rich enough multipath so that the number of multipath components is not a concern.

The following two requirements are evident from the formulation of channel capacity in (2-3).

3. The link must employ no conventional mechanism, such as frequency- or code- division multiplexing, to ensure that the signals transmitted by different transmitting antennas are orthogonal to each other at the receiver.

4. The receiver must measure the channel gain (both amplitude gain and phase shift). The techniques introduced in this book are all based on an implicit assumption that the receiver can apply coherent processing techniques over the received signals.

2.6 Summary

In this chapter, we outlined the scope of study for the next few chapters. Specifically, we provided a general framework for dual antenna-array systems operating on frequency-nonselective, burst-stationary Rayleigh fading channels. We provided the definition of channel capacity, assuming that the CSI at the receiver is perfect. The expressions for channel capacity subject to various constraints on the second-order statistics of the input are presented.

We showed that when the input covariance matrix is constrained to be a diagonal matrix with identical diagonal entries, the channel capacity of an (n, n) channel per antenna converges almost surely to a constant determined by average SNR. We also showed that if the receiver does not have CSI, increasing the number of transmit antenna elements over the channel coherence time does not improve channel capacity. Finally we commented on the requirements that must be met in order to utilize this high capacity.

Appendix

If m and n tend to infinity in such a way that n/m tends to a limit y $\in [0, 1]$, then the largest eigenvalue of $HH^{\dagger}/m$, denoted by $\lambda_{\max}$, converges to $\lambda_{max} \to (1 + \sqrt{y})^2$ almost surely [12]. Therefore, if $n/m \to 0$ asymptotically, i.e. the increase of the number of antenna elements at the transmitting end of a link is insignificant compared to that at the receiving end, from (2-5), the channel capacity per receiving antenna converges to

$$\lim_{m \to \infty} \frac{C}{m} = \lim_{m \to \infty} \frac{1}{m} \log_2 \left[\det \left(I + \frac{\rho}{n} HH^{\dagger} \right) \right]$$

$$= \lim_{m \to \infty} \frac{1}{m} \sum_{k=1}^{n} \log_2 \left(1 + \frac{\rho}{n} \varepsilon_k^2 \right)$$

$$\leq \lim_{m \to \infty} \frac{n}{m} \log_2 \left(1 + \frac{m}{n} \right) + \lim_{m \to \infty} \frac{1}{m} \sum_{k=1}^{n} \log_2 \left(1 + \frac{\rho}{m} \varepsilon_k^2 \right) \qquad (2\text{-}11)$$

$$\leq 0 + \lim_{m \to \infty} \frac{n}{m} \log_2 (1 + \rho \lambda_{max})$$

$$= 0.$$

3

SPATIAL FADING CORRELATION AND ITS EFFECTS ON CHANNEL CAPACITY

3.1 Introduction

In Chapter 2 it is shown that, if the fades between pairs of transmit-receive antenna elements are independent and identically Rayleigh, for a given transmitter power, the channel capacity per antenna of an (n, m) channel as $\min(n, m)$ grows toward infinity converges to a nonzero constant determined by the average SNR almost surely.

The aforementioned assumption of i.i.d. fading has been made in many previous works that explore the channel capacity of dual antenna-array systems; e.g. [3], [13], [14]. However, in real propagation environments, the fades are not independent due, for example, to insufficient spacing between antenna elements. Different channel correlation profiles lead to different channel capacity distribution. It has been observed [9] that when the fades are correlated the channel capacity can be significantly smaller than when the fades are i.i.d. The goal of this chapter is to investigate through analytical methods the effects of fading correlations on dual antenna-array systems. To do this, we first need to quantify the spatial fading correlation for the particular class of fading channels of interest.

There have been many works that study the characteristics of spatial fading correlation, mainly motivated by the need to quantify the effect of spatial fading correlation on the performance of diversity reception systems ($n = 1$, $m > 1$). One approach is to record a large number of typical channel realizations through field measurements or through ray-tracing simulations, e.g., [9], [15] - [18]. Another approach is to construct a scatterer model that can provide a reasonable description of the scattering environments for the wireless application of interest. The advantage of using abstract models is that with a simple and intuitive model the essential characteristics of the channel can be clearly illuminated, and the insights obtained from the model can then be utilized in planning the detailed measurements and/or simulations. For an overview of the numerous scattering models, see [19]. Examples of the abstract model approach include [7], [15], [20] - [22]. It must be noted, however, that abstract models need to be validated. To our knowledge, the modeling of fading correlation and its effect on capacity when both the transmitter and receiver employ multiple antenna elements have not been addressed by previous works[1].

In this chapter, to model multipath propagation and fading correlation, we extend the "one-ring" model first employed by Jakes [20]. This model is appropriate in the context where the one end of the wireless link is elevated and seldom obstructed. It must therefore be noted that the results obtained in this chapter are not necessarily applicable to other classes of fading channels. From the "one-ring" model, the spatial fading correlation of a narrowband flat fading channel can be determined from the physical parameters of the model, which include antenna spacing, antenna arrangement, angle spread, and angle of arrival. In this chapter we will only apply the channel capacity distribution given the spatial fading correlation. The spatial fading correlation can also be applied in research areas related to other applications of multiple antenna systems [24].

---

1. Driessen and Foschini [23] studied the deterministic channel when only line-of-sight channel components exist between the transmitting antenna elements or their images and the receiving antenna elements.

As mentioned above, in order to quantify the effect of fading correlation, we focus on the information-theoretic channel capacity. To interpret the effect of spatial fading correlation, we will first show that an (n, m) MIMO channel can be decomposed into $\min(n, m)$ subchannels, or (spatial) eigenmodes. The channel capacity of an (n, m) MIMO channel is the sum of the capacities of its individual subchannels. In Chapter 3.3 we show that spatial fading correlation determines the distributions of the subchannel capacities and thus the distribution of the overall channel capacity. We formulate closed-form expressions for the upper and lower bounds of the channel capacity and present the distributions of these bounds. The exact distributions of the overall channel capacity and subchannel capacities are difficult to compute, however; we employ Monte-Carlo simulations to observe histograms of these quantities.

This chapter is organized as follows. In Chapter 3.2, we present the abstract multipath propagation model from which the spatial fading correlation is derived. In Chapter 3.3, we present the analysis of channel capacity, most importantly the closed-form expressions for the distributions of the bounds on channel capacity given the spatial fading correlation. In Chapter 3.4 we employ Monte-Carlo simulations to obtain the histograms of channel capacity. Concluding remarks can be found in Chapter 3.6.

3.2 Scatterer Model and Spatial Fading Correlation

Fig. 3-1 shows the "one-ring" model. This model will be employed to determine the spatial fading correlation of the channel H. As we mentioned in the introduction, this model has been employed in several studies with some minor variations. The "one-ring" model is appropriate in the fixed wireless communication context, where the base station (BS) is usually elevated and unobstructed by local scatterers and the subscriber unit (SU) is often surrounded by local scatterers. For notational convenience, in this chapter the BS and the SU assume the roles of transmitter and receiver, respectively. In other words, we are taking a forward-link perspective. The parameters in the model include the distance D between BS and SU, the radius R of the scatterer ring, the angle of arrival Θ at the BS, and the geometrical arrangement of the antenna sets. As seen by a particular antenna element, the angles of incoming waves are confined within $[\Theta - \Delta, \Theta + \Delta]$. We refer to Δ as the angle spread. Since D and R are typically large compared to the antenna spacing,

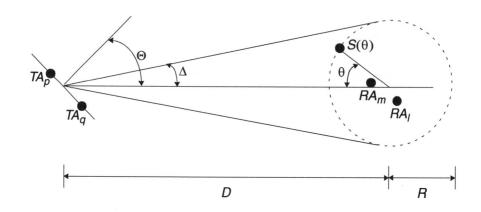

TA_p: transmitting antenna p
RA_l: receiving antenna l
S(θ): the scatterer at angle θ
Θ: angle of arrival
Δ: angle spread,
$D_{X \to Y}$: the distance from object X to object Y.

Fig. 3-1. Illustration of the abstract "one-ring" model. The size of the antenna sets are exaggerated for clarity.

$\Delta \approx \arcsin(R/D)$. The "one-ring" model is basically a ray-tracing model. The following assumptions are generally made in this model ([15], [21]):

- Every "actual scatterer" that lies at an angle θ to the receiver is represented by a corresponding "effective scatterer" located at the same angle on the scatterer ring centered on the SU. Actual scatterers, and thus effective scatterers, are assumed to be distributed uniformly in θ. The effective scatterer located at angle θ is denoted by S(θ). A phase $\phi(\theta)$ is associated with S(θ); $\phi(\theta)$ represents the dielectric properties and the radial displacement from the scatterer ring of the actual scatterer that S(θ) represents [15]. Therefore, rays that are reflected by S(θ) are all subject to a phase change of $\phi(\theta)$. Statistically, $\phi(\theta)$ is modeled as uniformly distributed in $[\pi, \pi)$ and i.i.d. in θ. The radius R of the scatterer ring is determined by the r.m.s. delay spread of the channel [15].

- Only rays that are reflected by the effective scatterers exactly once are considered.

- All rays that reach the receiving antennas are equal in power.

In the limit of infinitely many scatterers, the normalized complex path gain H_p^l from transmitting antenna element TA_p to receiving antenna element RA_l is

$$H_p^l = \frac{1}{\sqrt{2\pi}} \int_0^{2\pi} \exp\left\{-j\frac{2\pi}{\lambda}(D_{TA_p \to S(\theta)} + D_{S(\theta) \to RA_l}) + j\phi(\theta)\right\} d\theta. \quad (3\text{-}1)$$

In (3-1), $D_{X \to Y}$ is the distance from object X to object Y and λ is the wavelength. By the central limit theorem, H_p^l constructed from (3-1) is $\tilde{N}(0, 1)$. Therefore, in the limit case the channel constructed according to the model is purely Rayleigh fading [6] [25].

To study the spatial fading correlation, we use the following notation. If H is an $m \times n$ matrix then we use $\text{vec}(H)$ to denote the $mn \times 1$ vector formed by stacking the columns of H under each other; that is, if $H = (h_1 \; h_2 \; ... \; h_n)$, where h_i is an $m \times 1$ vector for $i = 1, ..., n$, then

$$\text{vec}(H) = (h_1', h_2', ..., h_n')'. \quad (3\text{-}2)$$

The covariance matrix of H is defined as the covariance matrix of the vector $\text{vec}(H)$: $\text{cov}(\text{vec}(H)) = E[\text{vec}(H)\text{vec}(H)^\dagger]$. (Note that for a zero-mean complex Gaussian vector g, the autocovariance is specified as the autocovariance matrix of the vector $(\text{Re}(g)' \; \text{Im}(g)')'$. Here, because it can be verified that $\text{vec}(H)$ constructed from the "one-ring" model is special complex Gaussian, the second-order statistics of $\text{vec}(H)$ are completely specified by $\text{cov}(\text{vec}(H))$ [2].) The covariance between H_p^l and H_q^k is

$$E[H_p^l H_q^{k*}] = \frac{1}{2\pi} \int_0^{2\pi} \exp\left\{\frac{-2\pi j}{\lambda}[D_{TA_p \to S(\theta)} - D_{TA_q \to S(\theta)}\right. $$

$$\left. + D_{S(\theta) \to RA_l} - D_{S(\theta) \to RA_k}]\right\} d\theta. \quad (3\text{-}3)$$

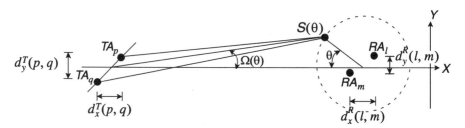

Fig. 3-2. Parameters used to derive the approximations for $E[H_p^l H_q^{k*}]$ in the "one-ring" model.

In general, (3-3) needs to be evaluated numerically. Fortunately, when Δ is small, which is often the case in fixed wireless applications, an approximation for (3-3) exists that offers useful insights. The approximation is derived using the notation illustrated in Fig. 3-2. In a two-dimensional plane, let the x-axis be parallel to the line that connects the BS and the SU. Let $d^T(p, q)$ denote the displacement between TA_p and TA_q, and $d_x^T(p, q)$ and $d_y^T(p, q)$ denote the projections of $d^T(p, q)$ on the x-axis and y-axis, respectively. Similar notations, $d^R(l, k)$, $d_x^R(l, k)$ and $d_y^R(l, k)$, apply to the SU side. Let Ω_θ denote the angle at which $S(\theta)$ is situated, as viewed from the center of the BS antenna relative to the x-axis. When Δ is small:

- $D_{TA_p \to S(\theta)} - D_{TA_q \to S(\theta)} \approx d_x^T(p, q) \cos \Omega_\theta + d_y^T(p, q) \sin \Omega_\theta$,

- $\sin \Omega_\theta \approx (R/D) \sin \theta \approx \Delta \sin \theta$,

- and $\cos \Omega_\theta \approx 1 - \frac{1}{2} \left(\frac{R}{D} \right)^2 \sin^2 \theta = 1 - \frac{1}{4} \left(\frac{R}{D} \right)^2 + \frac{1}{4} \left(\frac{R}{D} \right)^2 \cos 2\theta$.

Substituting these approximations into (3-3):

$$E[H_p^l H_q^{k*}] = \frac{1}{2\pi} \int_0^{2\pi} \exp\left\{ \frac{-2\pi j}{\lambda} [D_{TA_p \to S(\theta)} - D_{TA_q \to S(\theta)} \right.$$

$$\left. + D_{S(\theta) \to RA_l} - D_{S(\theta) \to RA_k}] \right\} d\theta$$

$$\approx \frac{1}{2\pi}\int_0^{2\pi}\exp\left\{-j\frac{2\pi}{\lambda}\left[d_x^T(p,q)\left(1-\frac{\Delta^2}{4}+\frac{\Delta^2\cos 2\theta}{4}\right)+\Delta d_y^T(p,q)\sin\theta\right.\right.$$

$$(3\text{-}4)$$

$$\left.\left.+ d_x^R(l,k)\sin\theta + d_y^R(l,k)\cos\theta\right]\right\}d\theta.$$

We evaluate (3-4) for the following special cases. Note that $(1/2\pi)\int_0^{2\pi}\exp(jx\cos\theta)d\theta = J_0(x)$, where $J_0(x)$ is the Bessel function of the first kind of the zeroth order.

- From one BS antenna element to two SU antenna elements, as $d^R(l,k)/R \to 0$, $E[H_p^l H_p^{k*}] \to J_0((2\pi/\lambda)d^R(l,k))$.
- From two BS antenna elements aligned on the y-axis to one SU antenna element, $d_x^T(p,q) = 0$, $E[H_p^k H_q^{k*}] \approx J_0(\Delta(2\pi/\lambda)d_y^T(p,q))$.
- From two BS antenna elements aligned on the x-axis to one SU antenna element, $d_y^T(p,q) = 0$,

$$E[H_p^k H_q^{k*}] \approx e^{-j\frac{2\pi}{\lambda}d_x^T(p,q)\left(1-\frac{1}{4}\Delta^2\right)}J_0\left(\left(\frac{\Delta}{2}\right)^2\frac{2\pi}{\lambda}d_x^T(p,q)\right).$$

A well-known result for diversity reception systems derived in [21] states that when maximal-ratio combining is employed the degradation in capacity is small even with fading correlation coefficients as high as 0.5. From our numerical evaluations, we find that this is also a good rule of thumb for the capacity of dual antenna-array systems (see Chapter 3.4). Here, to attain a correlation coefficient lower than 0.5, the minimum antenna element separations employed by the three cases are 0.24λ, $0.24\Delta^{-1}\lambda$, and $0.96\Delta^{-2}\lambda$, respectively.

If the minimum SU antenna spacing is sufficiently greater than half wavelength, the correlation introduced by finite SU antenna element spacing is low enough that the fades associated with two different SU antenna elements can be considered independent. Mathematically, if the SU antenna spacing is large enough, the n rows of H can be regarded as i.i.d. complex Gaussian row vectors with covariance matrix Ψ, where $\Psi_{p,q} = E[H_p^k H_q^{k*}]$. The channel covariance matrix in this case is[2] $\text{cov}(\text{vec}(H)) = \Psi \otimes I_m$. Similarly, if the SU and BS switches their roles as the transmitter and the receiver, $\text{cov}(\text{vec}(H)) = I_m \otimes \Psi$. Note that if $\text{cov}(\text{vec}(H)) = \Psi^R \otimes \Psi^T$, the statistical properties of H are identical to those of the product matrix $AH_w B^\dagger$ where H_w contains i.i.d. $\tilde{N}(0, 1)$ entries, $AA^\dagger = \Psi^T$, and $(BB^\dagger)' = \Psi^R$. In summary, if the fades experienced by different SU antenna elements can be considered independent, the following approximations can be used to analyze the channel capacity:

$H \sim H_w B^\dagger$ in the downlink (BS to SU) and

$H \sim AH_w$ in the uplink (SU to BS). $\qquad$ (3-5)

In (3-5), the notation $x \sim y$ means that "the distribution of x is identical to the distribution of y". We will verify in Chapter 3.4 that (3-5) is a good approximation in the sense that the distribution of the eigenvalues of $HH^\dagger$ – hence the channel capacity distribution – is closely approximated.

3.3 Analysis of Channel Capacity

The channel capacity of an (n, m) channel given the channel realization H subject to an average transmitter power constraint is described by (2-4):

$$C = \max_{tr(\Sigma_s) \le \rho\sigma_v^2} \log_2\left[\det\left(I + \frac{1}{\sigma_v^2}H\Sigma_s H^\dagger\right)\right] \text{ (bits per channel use).} \quad (3-6)$$

Without loss of generality, in this chapter the noise variance σ_v^2 is set to 1.

---

2. The Kronecker product of matrices M and N is defined as

$$M \otimes N = \begin{bmatrix} M(1, 1)N & M(1, 2)N & \dots \\ M(2, 1)N & M(2, 2)N & \dots \\ \dots & \dots & \dots \end{bmatrix}.$$

The MIMO channel is an n-input, m-output linear channel with i.i.d. AWGN. With linear operations at both the input and the output of the channel, an (n, m) channel can be transformed into an equivalent system consisting of $\min(n, m)$ decoupled single-input, single-output (SISO) subchannels. To show this, let the singular value decomposition of the channel matrix H be $H = U_H D_H V_H^\dagger$. The transmitter left-multiplies the signal to be conveyed, x_τ, by the unitary matrix V_H. Similarly, the receiver left-multiplies the received signal r_τ by $U_H^\dagger$. That is, $s_\tau = V_H x_\tau$, $y_\tau = U_H^\dagger r_\tau$, and $u_\tau = U_H^\dagger v_\tau$. These unitary transforms do not affect the channel capacity. Substituting these into (2-2), the input-output relationship between x_τ and y_τ is

$$y_\tau = D_H x_\tau + u_\tau, \qquad (3\text{-}7)$$

where the components of the noise vector u_τ are i.i.d. $\tilde{N}(0, 1)$. Denote the diagonal entries of the nonnegative diagonal matrix D_H by ε_k, $k = 1, 2, \ldots, n$. Writing (3-7) component-wise, we get

$$y_\tau^k = \varepsilon_k x_\tau^k + u_\tau^k, \ k = 1, 2, \ldots, n. \qquad (3\text{-}8)$$

Therefore, the multiplication of unitary matrices V_H and $U_H^\dagger$ transforms an (n, m) MIMO channel into n SISO subchannels with (power) gains ε_k^2. Note that ε_k^2 are the eigenvalues of $HH^\dagger$ because $HH^\dagger U_H = U_H D_H^2$. The channel capacity is the sum of the capacities of the n subchannels [2]. Suppose that a normalized transmit power ρ_k is allocated to the kth subchannels, the channel capacity is

$$C = \sum_{k=1}^{n} \log_2(1 + \rho_k \varepsilon_k^2). \qquad (3\text{-}9)$$

From (3-9), channel capacity is determined by both and ε_k^2, which is a function of H, and ρ_k, which does not depend on H. In this chapter, the focus is on the effect of channel fading correlation. Not to make things overly complicated, in this chapter we assume that the transmitted power is distributed evenly to these subchannels [26]; i.e. $\rho_k = \rho/n$. This is referred to as uniform power allocation. Uniform power allocation is robust, easy to analyze, and amenable to implementation ([9], [26], [27]). We will examine the general problem of power allocation in further detail in the next chapter.

The channel capacity subject to a uniform power allocation constraint $\rho_k = \rho/n$ is

$$C = \log_2\left(\det\left(I + \frac{\rho}{n}HH^\dagger\right)\right) = \sum_{k=1}^n \log_2\left(1 + \frac{\rho}{n}\varepsilon_k^2\right). \qquad (3\text{-}10)$$

Note that in this case the channel capacity is independent of V_H. This property makes uniform power allocation a good choice for systems in which the transmitter cannot acquire the knowledge of H.

3.3.1 Bounds on Channel Capacity

The distribution of channel capacity can be calculated given the distribution of ε_k^2. However, for a general spatial fading covariance and a finite spatial dimensionality, the distribution of ε_k^2 can be very difficult to compute. The exact distributions of ε_k^2 and channel capacity will be studied using Monte-Carlo simulations in the next section. Here, we formulate lower and upper bounds on channel capacity based on the fading statistics (3-5). To derive these bounds, we need the following mathematical tools.

(a) Let H_w be an $m \times n$ matrix whose entries are i.i.d. $\tilde{N}(0, 1)$. The subscript w is used to mean "white". Denote the QR decomposition of H_w by $H_w = QR$, where Q is an orthogonal matrix and R is an upper triangular matrix. The upper diagonal entries of R are i.i.d. $\tilde{N}(0, 1)$ and are statistically independent of each other. The magnitude squares of the diagonal entries of R, say $|R_l^l|^2$, are chi-squared distributed with $2(m - l + 1)$ degrees of freedom. These can be proved by applying the standard Householder transformation to the matrix H_w [28], [29]. Clearly, $H_w Q_1 \sim Q_2 H_w \sim H_w$ for any unitary matrices Q_1 and Q_2.

(b) For any diagonal matrix D and any upper-triangular matrix R,
$\det(DD^\dagger + RR^\dagger) \geq \prod_l (|D_l^l|^2 + |R_l^l|^2)$.

(c) For any nonnegative definite matrix A, $\det(A) \leq \prod_l A_l^l$.

(d) For any unitary matrix Q and any square matrices X and Y,
$\det(I + XY) = \det(I + YX)$ and $\det(I + QXQ^\dagger) = \det(I + X)$.

Next, we examine the following two special cases. In the following, we assume that $n \leq m$.

Case I. The fades are i.i.d. Substituting $H_w = QR$ into (3-10), the channel capacity can be lower- and upper-bounded by, respectively,

$$C = \log_2\left(\det\left(I + \frac{\rho}{n}HH^\dagger\right)\right) \overset{(d)}{=} \log_2\left(\det\left(I + \frac{\rho}{n}RR^\dagger\right)\right)$$

$$\overset{(b)}{\geq} \sum_{l=1}^{n} \log_2\left(1 + \frac{\rho}{n}|R_l^l|^2\right),\tag{3-11}$$

and

$$C = \log_2\left(\det\left(I + \frac{\rho}{n}RR^\dagger\right)\right)$$

$$\overset{(c)}{\leq} \sum_{l=1}^{n} \log_2\left(1 + \frac{\rho}{n}(|R_l^l|^2 + \sum_{k=l+1}^{n}|R_k^l|^2)\right).\tag{3-12}$$

From (a), $|R_l^l|^2$ is chi-squared distributed with $2(m-l+1)$ degrees of freedom. Also because $\sum_{k=l+1}^{n}|R_k^l|^2$ is chi-squared distributed with $2(n-l)$ degrees of freedom, the term $|R_l^l|^2 + \sum_{k=l+1}^{n}|R_k^l|^2$ in (3-12) is chi-squared distributed with $2(m+n-2l+1)$ degrees of freedom. In short, the channel capacity is lower bounded by the sum of the capacities of n subchannels whose power gains are independently chi-squared distributed with degrees of freedom $2m$, $2m-2$, ..., $2(m-n+1)$, and is upper bounded by the sum of the capacities of n subchannels whose power gains are independently chi-squared distributed with degrees of freedom $2(m+n-1)$, $2(m+n-3)$, ..., $2(m-n+1)$. The difference between the mean values of the upper and the lower bounds is no greater than 1 bps/Hz per spatial dimension. The lower bound was first derived by Foschini in [26]. In fact, Foschini has proved that the mean values of the exact channel capacity and its lower bound, both normalized to per-spatial dimension quantities, converge to the same limit when $n \to \infty$ [26].

Case II. $\text{cov}(\text{vec}(H)) = \Psi \otimes I_m$ or $I_m \otimes \Psi$. We have shown in Chapter 3.2 that in the "one-ring" model if the antenna array inside the scatterer ring (usually the SU) employs a sufficiently large antenna element spacing, the fading covariance matrix can be approximated by $\Psi \otimes I_m$ in the downlink (BS to SU) and $I_m \otimes \Psi$ in the uplink (SU to BS), and the approximations in (3-5) apply. Note that if $\text{cov}(\text{vec}(H)) = \Psi \otimes I_m$ for some nonnegative definite Ψ, the distributions of ε_k^2 and hence the distribution of channel capacity can be exactly calculated using the techniques developed for Wishart matrices [28]. However, the calculation is generally very difficult because it involves the zonal polynomials, which are notoriously difficult to compute. Furthermore, the actual computation does not give us as much insight into the problem compared to the following bounds.

Substituting H by $AH_wB^\dagger$ into (3-10), we have

$$C \sim \log_2\left[\det\left(I + \frac{\rho}{n}AH_wB^\dagger(AH_wB^\dagger)^\dagger \right) \right]$$

$$\overset{(a)(d)}{\sim} \log_2\left[\det\left(I + \frac{\rho}{n}H_wD_B^2H_w{}^\dagger D_A^2 \right) \right]. \tag{3-13}$$

Here D_A and D_B are diagonal matrices whose diagonal elements are the singular values of A and $B^\dagger$, respectively. The diagonal entries of both D_A and D_B are ordered in descending order of their magnitudes down the diagonal. Substituting $H_w = QR$ and $A = I$ in (3-13), the capacity in the downlink can be bounded by:

$$C \sim \log_2\left[\det\left(I + \frac{\rho}{n}H_wD_B^2H_w{}^\dagger \right) \right]$$

$$\overset{(b)(d)}{\geq} \sum_{l=1}^n \log_2\left(1 + \left(\frac{\rho}{n}\right)|D_B(l,l)|^2|R_l^l|^2 \right) \tag{3-14}$$

and

$$C \overset{(c)(d)}{\leq} \sum_{l=1}^n \log_2\left(1 + \left(\frac{\rho}{n}\right)|D_B(l,l)|^2(|R_l^l|^2 + \sum_{k=l+1}^n |R_k^l|^2) \right). \tag{3-15}$$

Similar to the case when the fades are i.i.d., the channel capacity is still lower- and upper- bounded by the total capacity of n independent SISO subchannels, and the difference between the mean values of the upper and the lower bound is less than 1 bps/Hz per spatial dimension. Due to the spatial fading correlation, the power gain of the lth subchannel is scaled by a factor of $|D_B(l, l)|^2$ (or, on decibel scale, augmented by $10\log_{10}|D_B(l, l)|^2$ dB). Note that because the trace of D_B^2 is equal to n, when compared to the situation in which the fades are i.i.d., the path gains of some subchannels are scaled up while others are scaled down.

When the number of antenna elements is large, determining the channel capacity through simulation is very computation-intensive. The upper bound in (3-15) can be employed to investigate the capacity when the number of antenna elements is large. Let $E(C)$ denote the mean value of channel capacity at a fixed average total power constraint ρ. For any concave function $f(x)$, $E(f(x)) \le f(E(x))$. Thus an upper bound of $E(C)$, denoted by $\bar{E}(C)$, in the downlink direction can be derived from (3-15) by substituting the mean values of chi-squared random variables for them:

$$E(C) \le \sum_{l=1}^{n} \log_2\left(1 + \frac{\rho}{n}|D_B(l, l)|^2 E(|R_l^l|^2 + \sum_{k=l+1}^{n}|R_k^l|^2)\right)$$

$$= \sum_{l=1}^{n} \log_2\left(1 + \frac{\rho}{n}|D_B(l, l)|^2(m + n - 2l + 1)\right) \equiv \bar{E}(C). \tag{3-16}$$

Note that due to the normalization used in this chapter, the mean value of a chi-squared random variable with $2k$ degrees of freedom is k.

For an example of the applications of the bounds, we employ (3-16) to investigate the effect of angle spread on the relation between $\bar{E}(C)$ and the number of antenna elements $n = m$. The result is displayed in Fig. 3-3. (The definitions of broadside and inline linear antenna arrays will be given in Chapter 3.4.) The bounds permit us to compute $\bar{E}(C)$ even when n is large.

3.3.2 Effective Degrees of Freedom

We have shown in (3-9) that an (n, m) channel can be decomposed into an equivalent system of $\min(n, m)$ SISO subchannels whose path power gains are the eigenvalues of $HH^\dagger$. Based on this decomposition, one would intuitively

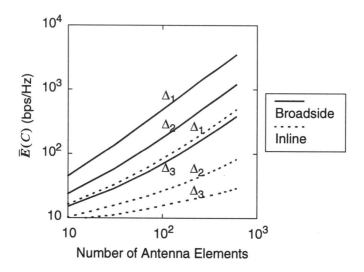

Fig. 3-3. The effect of angle spread Δ on the relationship between the upper bound of mean capacity, $\bar{E}(C)$, and the number of antenna elements $n = m$. The fixed overall power constraint is $\rho = 18\text{dB}$.

expect that the channel capacity of an (n, n) channel grows roughly linearly with n for a given fixed transmitted power, because if $\rho\varepsilon_k^2/n \gg 1$ for $k = 1, 2, ..., n$, (3-10) can be approximated by

$$C(\rho) \approx \sum_{k=1}^{n} \log_2\left(\frac{\rho\varepsilon_k^2}{n}\right). \tag{3-17}$$

However, this high-SNR condition may not be met in practice. If $\rho\varepsilon_k^2/n$ is much smaller than one for some k, the capacity provided by the kth sub-channel is nearly zero. This may occur when the communication system operates in a low-SNR setting, e. g., in long-range communication application or transmission from low-power devices. On the other hand, it may occur if with significant probability ε_k^2 is very small, which is a direct result of severe fading correlation. Here we introduce the concept of effective degrees of freedom (EDOF), which is a parameter that represents the number of sub-channels actively participating in conveying information under a given set of operating conditions. It is well known that for an SISO channel, at high SNR a G-fold increase in the transmitter power results in an increase in the channel capacity of $\log_2 G$ bps/Hz. If a system is equivalent to EDOF SISO channels in parallel, the capacity of the system should increase by $(\text{EDOF} \cdot \log_2 G)$ bps/Hz when the transmitter power is raised by a factor of G. In light of this, we define EDOF at a given transmit power ρ and outage probability q to be

$$\text{EDOF} \equiv \frac{d}{d\delta} C_q(2^\delta \rho)\Big|_{\delta = 0}. \tag{3-18}$$

We note that EDOF is a real number in $[0, n]$. Although the $m \times n$ channel matrix H has rank n with probability one in general, the power allocated to $(n - \text{EDOF})$ out of the n dimensions is very poorly utilized. EDOF is a function of spatial fading correlation and SNR; its value is higher when SNR is increased.

For an extreme example of how fading correlation affects EDOF, consider the fading correlation in the "one-ring" model when the angle spread approaches zero. In such a case, $\left|E[h_p^l h_q^{l*}]\right| \to 1$. Therefore, the n columns of H are perfectly correlated, and only one of the n eigenvalues of $HH^\dagger$ has significant probability of being practically nonzero. The overall effect is that, as the angle spread approaches zero, EDOF approaches one. The capacity that an (n, n) dual antenna-array system provides thus degenerates to that provided by a $(1, n)$ multiple antenna system.

3.4 Simulation Results

In this section, we present the capacity of dual antenna-array systems obtained from Monte-Carlo simulations. Simulation is necessary because computing the distributions of channel capacity, subchannel gains and subchannel capacities analytically is very difficult. The results in this section illustrate the effect of the antenna geometry and the physical dimensions of the scattering environment on the statistics of channel capacity. Another goal is to verify that (3-5) is a good approximation to the exact channel distribution.

3.4.1 Simulation Algorithm

Fig. 3-4 shows the arrangement of antenna elements. We have chosen a fixed number of antenna elements $m = n = 7$. In Fig. 3-4(a), seven antenna elements are equally spaced along an axis. This is referred to as a linear antenna array. In Fig. 3-4(a) we also define the angle of arrival Θ at the BS for linear antenna arrays. Following conventional notation [15], we use the term "broadside" and "inline" to refer to the situations when $\Theta = 0°$ and $\Theta = 90°$, respectively. In Fig. 3-4(b), seven antennas are arranged on a hexagonal planar array. This is referred to as the hexagon antenna array. For planar antenna sets, the hexagonal arrangement achieves the highest antenna density per unit area for a given nearest-neighbor antenna spacing. Furthermore, the effects of the angle of arrival are not significant, due to the symmetry of the hexagon. Three configurations are considered: broadside and inline linear antenna array at the BS with inline linear antenna array at the SU, and hexagon antenna arrays at both the BS and the SU. The nearest-neighbor separations between antenna elements of the BS and the SU antenna sets are denoted by dt and dr, respectively. Again the BS and the SU assume the roles of the transmitter and the receiver, respectively.

Given Δ, dt, and dr, one way to generate the channel realization is to randomly select the angular positions and phases of the equivalent scatterers and compute H using ray tracing. When the number of scatterers is large, an equivalent way is as follows. First, compute the channel covariance matrix $\text{cov}(\text{vec}(H))$ from (3-4). Let $\Psi = \text{cov}(\text{vec}(H))$ and $\Psi = \Psi^{1/2}(\Psi^{1/2})^{\dagger}$. The

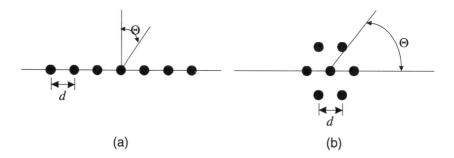

(a)   (b)

Fig. 3-4. Antenna array arrangement. Note the definitions of the angle of arrival Θ and the minimum antenna element spacing d. (a) Linear antenna array. (b) Hexagon antenna array.

instances of H can then be generated by premultiplying a white channel, $\text{vec}(H_w)$, by $\Psi^{1/2}$. That is,

$$\text{vec}(H) = \Psi^{1/2}\text{vec}(H_w). \tag{3-19}$$

We generated 10,000 instances of channel and collected the statistics of channel capacity and ordered eigenvalues of $HH^\dagger$. The average received SNR ρ is chosen to be 18 dB. For comparison purposes, the 10% outage channel capacities $C_{0.1}$ of $(1, 1)$, $(1, 7)$, and $(7, 7)$ systems over i.i.d. Rayleigh-fading channels given $\rho = 18$ dB are 2.94, 7.99, and 32.0 bps/Hz, respectively.

3.4.2 Results

The physical parameters in the "one-ring" model include the angle spread, angle of arrival, antenna spacing, and antenna arrangement. First, we investigate the effect of angle spread Δ. Fig. 3-5(a) shows the complementary cumulative distribution function (ccdf) of channel capacity with hexagon antenna arrays versus Δ. The support of the transition region of the ccdf curve moves toward lower capacity values as the angle spread decreases. Note that when the angle spread is extremely small ($\Delta < 0.6°$), the ccdf for the channel capacity of a $(7, 7)$ dual antenna-array systems with hexagon antenna arrays is identical to that of a $(1, 7)$ diversity reception system with maximal-ratio combining. Fig. 3-5(b) shows $C_{0.1}$ for the three configurations of antenna arrays versus Δ. For all three, $C_{0.1}$ decreases monotonically as the angle spread decreases. Intuitively, because the difference in path lengths from two transmitting antenna elements to any scatterer becomes smaller as Δ decreases, it becomes increasingly difficult for the receiver to distinguish between the transmissions of the various transmitting antenna elements. Mathematically, the correlation between the columns of H increases as Δ decreases. Fig. 3-5(c) shows that the EDOF of each type of antenna array settings indeed decreases as the angle spread decreases.

The simulation also provides the pdfs of the ordered eigenvalues of $HH^\dagger$. The magnitude of ε_k^2 is best displayed in decibel units. Let $\mu_k = 10\log_{10}\varepsilon_k^2$, and let $p_k(\mu_k)$ denote the pdf of μ_k. Fig. 3-6 displays $p_k(\mu_k)$. The followings are observed. As the angle spread Δ decreases, (a) the median of μ_1 increases slightly, (b) the medians of μ_k, $k \geq 2$, decrease, and (c) the difference between

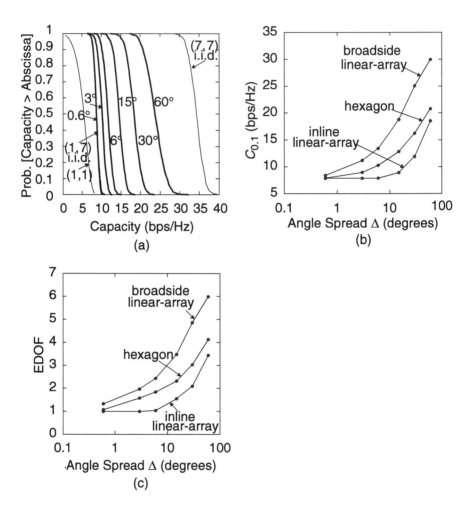

Fig. 3-5. (a) The ccdf of channel capacity with hexagon antenna arrays given various angle spreads. $dt = dr = 0.5\lambda$. The reference curves are the ccdfs of channel capacity when assuming H is a 7×7, 1×7, and 1×1 matrix with i.i.d. $\tilde{N}(0, 1)$ entries, respectively. (b) $C_{0.1}$ versus angle spread. (c) EDOF versus angle spread.

the medians of μ_k and μ_{k+1} increases for all k. These observations indicate that, statistically, as Δ decreases, the disparity among μ_k, i.e. the disparity

among the subchannels in (3-8), increases. The pdfs ε_k^2 also provide a conve-
nient way to estimate the EDOF. The average received SNR necessary to
obtain a certain EDOF can be estimated from Fig. 3-6 as follows. For a natural
number z, the average received SNR necessary for EDOF = z is approximately
$-\alpha$, where α is determined by $\int_\alpha^\infty p_z(\mu_z)d\mu_z = 0.9$.

Secondly, we investigate the effect of the BS antenna spacing dt.
Fig. 3-7(a) shows the ccdf of channel capacity with hexagon antenna arrays in
the large angle spread setting ($D = 1000\lambda$, $\Delta = 15°$, $dr = 0.5\lambda$). We find
that the channel capacity increases greatly as dt increases. In Fig. 3-7(a), similar to Fig. 3-5(a), the support of the transition part of the ccdf curve moves
toward higher capacity values as dt increases. Fig. 3-7(b) and Fig. 3-7(c) display the relation between $C_{0.1}$ and dt for the three types of antenna array settings in the large and small ($D = 100,000\lambda$, $\Delta = 0.6°$, $dr = 0.5\lambda$) angle
spread settings, respectively. Given a fixed dt, the capacity of a (7, 7) system
with broadside linear antenna array is always higher than that of a (7, 7)
system with hexagon antenna array which, in turn, is always higher than that
of a (7, 7) system with inline linear antenna array. In Chapter 3.3, we showed
that the effectiveness of reducing the fading correlation by increasing the BS
antenna spacing along the axes perpendicular and parallel to the arriving
waves are different. To attain zero fading correlation with inline linear antenna
arrays, the BS antenna spacing must be $4/\Delta$ times of the spacing required
when using broadside linear antenna arrays. The difference in effectiveness is
confirmed here by simulation. Note that because the Bessel function gov-
erning the relation between antenna spacing and fading correlation is not
monotonic, the channel capacity does not decrease monotonically as dt is
decreased. This can be seen in Fig. 3-7(b).

Thirdly, we examine the effect of the SU antenna spacing dr. Fig. 3-8(a)
shows the ccdf of channel capacity with hexagon antenna arrays in the large
angle spread setting ($D = 1000\lambda$, $\Delta = 15°$, $dt = 0.5\lambda$). Fig. 3-8(b) and
Fig. 3-8(c) display $C_{0.1}$ versus dr in the large and small ($D = 100,000\lambda$,
$\Delta = 0.6°$, $dt = 5\lambda$) angle spread settings, respectively. The ccdf curves of
channel capacity become steeper as dr increases. This results in an improvement in $C_{0.1}$, but such an improvement is not nearly as significant as the
capacity improvement while increasing dt. The analysis in Chapter 3.3
explains the striking difference between increasing the antenna spacing at the
BS and at the SU, in terms of effectiveness in improving channel capacity.
Once the antenna spacing at the SU is more than a half wavelength, the corre-

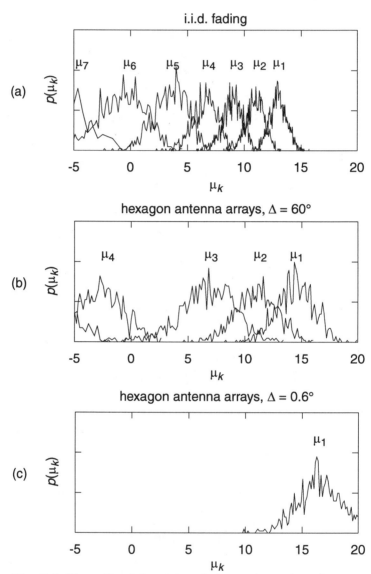

Fig. 3-6. The pdfs of the ordered eigenvalues of $HH^\dagger$ based on the "one-ring" model. Here, ε_k^2 is the kth largest eigenvalue of $HH^\dagger$ and $p(\mu_k)$ is the pdf of $\mu_k = 10\log_{10}\varepsilon_k^2$. The number of antennas is $n = m = 7$, and hexagon antenna arrays are employed. The pdfs are normalized to have the same height for display purpose. (a) I.i.d. fades. (b) Large angle spread setting. (c) Small angle spread setting.

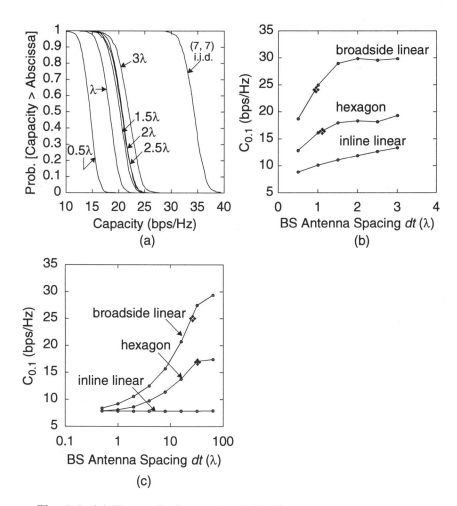

Fig. 3-7. (a) The ccdf of capacity of (7, 7) hexagon antenna array with $\Delta = 15°$, $dr = 0.5\lambda$. (b) $C_{0.1}$ versus dt for large angle spread ($\Delta = 15°$, $dr = 0.5\lambda$). (c) $C_{0.1}$ versus dt for small angle spread ($\Delta = 0.6°$, $dr = 0.5\lambda$). In (b) and (c) we use ❖ to mark the smallest dt with which the maximum fading correlation coefficient is 0.5. After the maximum fading correlation coefficient is reduced to under 0.5, the benefit of increasing dt starts to saturate.

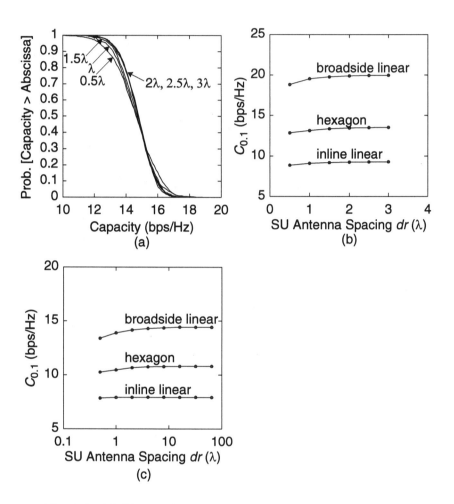

Fig. 3-8. (a) The ccdf of capacity of a (7, 7) dual antenna-array system with hexagon antenna arrays at various values of dr with $\Delta = 15°$, $dt = 0.5\lambda$. (b) $C_{0.1}$ versus dr for large angle spread ($\Delta = 15°$, $dt = 0.5\lambda$). (c) $C_{0.1}$ versus dr for small angle spread ($\Delta = 0.6°$, $dt = 5\lambda$).

lation coefficient between any two entries on a column of H is generally lower than 0.5. The fading correlation is already low and therefore cannot be reduced significantly by increasing dr.

We conclude that the angle spread and the BS antenna spacing perpendicular to the direction of the arriving waves at the BS dominates the channel correlation and thus the channel capacity. If the incoming waves are known to come from a certain direction, it is advantageous to deploy a broadside linear antenna array. On the other hand, if omnidirectional coverage is the goal, an antenna array with a symmetric configuration, such as the hexagon antenna array, is clearly the better choice.

Fig. 3-9 compares the eigenvalue distributions of $HH^\dagger$ and $(H_w B^\dagger)(H_w B^\dagger)^\dagger$ given the parameters $dt = dr = 3\lambda$ and $\Delta = 15°$ and $0.6°$. Very good agreement is observed. The results in Fig. 3-8 also show that the overestimate of channel capacity caused by assuming the rows of H are uncorrelated is not substantial. These results demonstrate that (3-5) is a valid approximation in the downlink if the SU employs an antenna spacing sufficiently large.

3.5 Two-ring Model

In certain applications such as indoor wireless systems and mobile-to-mobile communications, it is common to find that both ends of the link are surrounded by local scatterers. In these cases the "one-ring" model is no longer appropriate. Fig. 3-10 illustrates the "two-ring" model, which is a natural extension of the "one-ring" model. In the "two-ring" model, a communication entity always has a ring of scatterers centered around it.

The path gain $H_{l,p}$ in the "two-ring" model is obtained through ray-tracing in a manner similar to (3-1) in the "one-ring" model. That is, ignoring the path amplitude loss,

$$H_p^q = \frac{1}{\sqrt{K_1 K_2}} \sum_{k=1}^{K_1} \sum_{l=1}^{K_2} \exp\left\{ -j\frac{2\pi}{\lambda}(D_{TA_p \to S_1(\theta_k)} \right.$$
$$\left. + D_{S_1(\theta_k) \to S_2(\theta_l)} + D_{S_2(\theta_l) \to RA_q}) + j\phi_1(\theta_k) + j\phi_2(\theta_l) \right\}. \tag{3-20}$$

In (3-20), $S_u(\theta_k)$ refers to the kth scatterer on the scatterer ring around user u, $u = 1, 2$. The additive phase component $\phi_u(\theta_k)$ refers to the phase associated

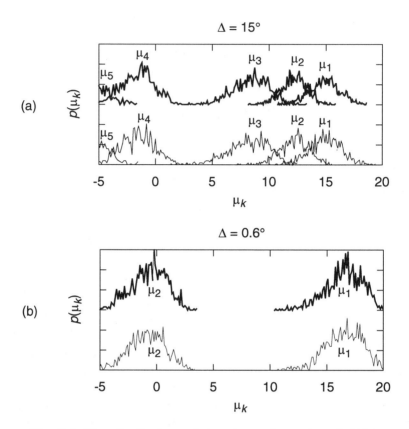

Fig. 3-9. The distributions of the eigenvalues of $HH^\dagger$ (thick, elevated curves) and $H_w B^\dagger (H_w B^\dagger)^\dagger$ (thin curves) given the parameters $dt = dr = 3\lambda$ and $\Delta = 15°$ (a) and $0.6°$ (b).

with the scatterer $S_u(\theta_k)$. It is assumed that the additive phases are independently uniformly distributed.

Unfortunately, in contrast to the "one-ring" model, as both $K_1 \to \infty$ and $K_2 \to \infty$, the channel gain H_p^q does not converge to a Gaussian random variable. Therefore, it is not enough to completely describe the statistics of the channel by specifying the channel covariance matrix $\text{cov}(\text{vec}(H))$. Instead, we will generate instances of channel realizations through ray-tracing and use them in subsequent Monte-Carlo simulations to study the statistical properties of the channel.

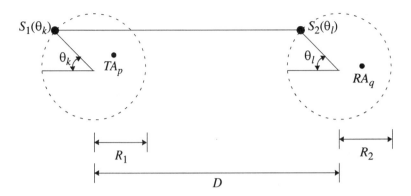

Fig. 3-10. Illustration of the abstract "two-ring" model. The size of the antenna sets are exaggerated for clarity.

Fig. 3-11 displays the capacity distribution at different separations between the two communicating entities. It is obvious that as the distance between the two communicating entities increases, the channel capacity decreases as a result of increasing fading correlation. Because the two antenna arrays are both surrounded by local scatterers, the variation of the capacity distributions due to the orientation and the physical configurations of the antenna arrays is insignificant. We also found that the effect of antenna spacing is small if the antenna spacing is more than half a wavelength.

3.6 Summary

In previous studies that analyze the channel capacity of dual antenna-array systems, a common assumption is that the fades between pairs of transmit-receive antenna elements are i.i.d. However, in real propagation environments, fading correlation does exist, and can potentially lead to a capacity lower than that predicted under the i.i.d. fading assumption. In this chapter, we proposed an abstract model for the multipath propagation environment. Using the model, the spatial fading correlation and its effect on the channel capacity can be determined.

The "one-ring" model can reasonably represent a scattering environment in which one of the communicating parties, the SU, is surrounded by local scatterers. The channel correlation based on the "one-ring" model is a func-

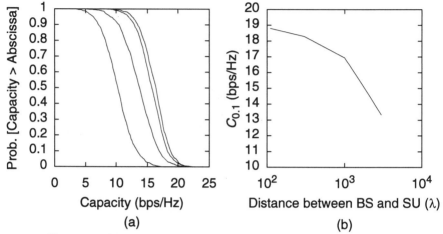

Fig. 3-11. Capacity according to the "two-ring" model. (a) The ccdf of capacity of a (7, 7) dual antenna-array system with linear antenna arrays at various distance between the communicating entities. (b) $C_{0.1}$ versus distance D.

tion of antenna spacing, antenna arrangement, angle spread, and the angle of arrival. When the angle spread is small, the contributions to the spatial fading correlation from the SU antenna element spacing and the BS antenna element spacing (both parallel to and perpendicular to the direction of wave arrival) are significantly different. We derived expressions for approximate fading correlation to highlight their differences. We considered the situations in which the antenna element spacing at the SU is sufficient that the correlation among the entries on any column of the channel matrix is negligible.

To understand the effect of fading correlation on channel capacity analytically, we first showed that an (n, m) MIMO channel consists of $\min(n, m)$ SISO subchannels, or eigenmodes. The MIMO channel capacity is the sum of the individual subchannel capacities; the gains of these subchannels are the $\min(n, m)$ largest eigenvalues of $HH^\dagger$. When the SU antenna spacing is sufficiently large, the power gains of these subchannels are independent scaled chi-squared distributed random variables with $2(m-n+1), 2(m-n+2), \ldots, 2m$ $(2(m-n+1), 2(m-n+3), \ldots, 2(m+n-1))$ degrees of freedom. The fading correlation determines the scaling factors. The stronger the fading correlation, the higher the disparity between these scaling factors. As the fading correlation becomes more severe, more and more subchannels have gains too small to convey information at any significant rate. We defined the parameter effective degrees of freedom (EDOF) to represent the number of subchannels that actively contribute to the overall channel capacity.

We performed Monte-Carlo simulations to study quantities that are very difficult to compute analytically, such as the distributions of the eigenvalues of $HH^{\dagger}$ and the channel capacity. We found that when the angle spread is small, the product of angle spread and antenna spacing perpendicular to the direction of wave arrival is a key parameter. In general, the higher the product, the higher the channel capacity. The BS antenna separation parallel to the direction of wave arrival has a much less importance in determining fading correlation unless the separation is very large. If the direction of wave arrival is known approximately, it is advantageous to deploy a broadside linear antenna array with a large antenna spacing; but if omnidirectional coverage is the goal, an antenna array with a symmetric configuration, such as a hexagon antenna array, is clearly the best choice.

4

POWER-ALLOCATION
STRATEGIES

4.1 Introduction

The channel capacity of an (n, m) channel H subject to the autocovariance constraint on the input signal $E[s_\tau s_\tau^\dagger] = \Sigma_s$ is

$$C = \log_2\left[\det\left(I + \frac{1}{\sigma_v^2}H\Sigma_s H^\dagger\right)\right] \text{ (bits per channel use).} \qquad (4\text{-}1)$$

In this chapter, a power-allocation strategy specifically refers to the way the autocovariance matrix Σ_s is chosen.

The objective of a power-allocation strategy is to achieve a high capacity given the power and channel knowledge available at the transmitter. Throughout this chapter, we assume that the total average transmit power (i.e., sum over all n antennas) is constrained. We consider three power-allocation strategies: the optimum power allocation, the uniform power allocation, and the stochastic water-filling power-allocation strategy. The choice of power-allocation strategy depends on the type of CSI available to the transmitter.

In this chapter, power-allocation strategies are regarded as constraints on the channel. The channel capacity with a certain power allocation strategy means the maximal achievable mutual information between the input and the output given that the autocovariance matrix Σ_s must be chosen according to the particular power-allocation strategy.

If instantaneous CSI is available to the transmitter, then the optimum power allocation can be employed [2], [9]. In many applications, however, practical difficulties make it impossible for the transmitter to have instantaneous CSI. We use the term "blind transmission" to describe situations in which the transmitter does not have instantaneous CSI. In blind transmission systems, optimum power allocation cannot be used, and uniform power allocation, which allocates equal power to each individual transmitting antenna, is usually considered, e.g., [26]. Besides being applicable to blind transmission systems, uniform power allocation is robust, easy to implement, and easy to analyze.

In some situations, such as fixed wireless systems, although the transmitter does not have access to instantaneous CSI, it can acquire knowledge of the spatial correlation properties of the channel fading. This is possible because the spatial fading correlation properties are locally stationary. For blind transmission systems in which the channel correlation is known, we propose the stochastic water-filling power allocation strategy. The stochastic water-filling power allocation is inspired by the conventional water-filling procedure, and is computed using a procedure described below.

A major goal of this chapter is to examine the combined effects of the power-allocation strategy and the fading correlation on the capacity of dual antenna-array systems. We have demonstrated in the previous chapter that in environments where spatial fading correlation is strong, this independent fading idealization often leads to a significant overestimation of channel capacity. In this chapter, the channel fading correlation is modeled using the "one-ring" model developed in Chapter 3. This model is appropriate for typical outdoor fixed wireless applications.

We compare the channel capacities obtained in independent- and correlated-fading environments with the three power allocation strategies. We will show that if the fades are independent, for medium to high SNR, optimum and uniform power-allocation strategies offer nearly equal capacities. Therefore, with low fading correlation, the availability of CSI at the transmitter does not constitute a significant advantage. In contrast, when the fades are highly correlated, the difference between capacities achieved by optimum and uniform power allocations is significant. This result motivates us to devise the stochastic water-filling procedure. While it might at first seem impossible to perform an optimization similar to water-filling without knowledge of the instantaneous CSI, we will demonstrate that with high probability the stochastic water-filling power-allocation strategy achieves a significantly higher capacity than the uniform power allocation in the downlink direction (transmission from the unobstructed end of the link). We will also prove that in the uplink direction (the opposite direction of downlink), the uniform power allocation achieves the highest average channel capacity. This asymmetry arises because in the "one-ring" scatterer model, one end of the link is unobstructed, while the other end of the link is surrounded by a ring of scatterers.

The processing power available at the receiver can also influence the choice of power-allocation strategy. For blind transmission systems, if the transmitted data rate is increased in proportion to n, ML processing the received signal in general leads to a complexity that increases exponentially with n [43]. One way to reduce the receiver complexity is to one-dimensionalize the multidimensional signal-processing task. In other words, the receiver first derives n signals from the received signal, and then processes these n signals independently. An example is the layered space-time (LST) architecture [26]. Note that when optimum power-allocation strategy is employed, the (n, m) MIMO channel is automatically decomposed into an equivalent system of $\min(n, m)$ parallel single-input, single-output (SISO) channels with i.i.d. noises [2], [9]. Therefore, one-dimensional (1-D) signal processing is a direct result of using optimum power-allocation strategy, and the reduction of receiver complexity comes without loss of capacity. However, with other power-allocation strategies, there can be a capacity penalty associated with one-dimensionalizing. In this chapter, we examine this capacity penalty for systems that use the uniform power allocation and the stochastic water-filling power allocation.

The remainder of this chapter is organized as follows. In Chapter 4.2, we present lower bounds on capacities that we use to analyze the effect of fading correlation on channel capacity. In Chapter 4.3, we examine blind transmission systems in more detail. We propose the stochastic water-filling procedure and demonstrate that such a nonuniform power allocation achieves a higher capacity than uniform power allocation in the downlink direction. We also prove that in the uplink direction uniform power allocation achieves the highest average capacity. In Chapter 4.4, we introduce the concept of one-dimensional processing and quantify the associated capacity penalty. In Chapter 4.5, we provide numerical evaluations of channel capacity for typical configurations. We present concluding remarks in Chapter 4.6.

4.2 Power-Allocation Strategies

4.2.1 Optimum Power-Allocation Strategy

If the transmitter knows H, it can select Σ_s to maximize the mutual information in (4-1). We believe that the problem was first studied by Teletar in [2]. For the benefit of the readers, we briefly summarize the results here. Consider that the only constraint is an average power constraint, i.e. $tr(\Sigma_s) \le \rho\sigma_v^2$. Given a channel realization H, the channel capacity is

$$C = \max_{tr(\Sigma_s) \le \rho\sigma_v^2} \log_2\left[\det\left(I + \frac{1}{\sigma_v^2}H\Sigma_s H^\dagger\right)\right] \text{ (bits per channel use)}. \quad (4\text{-}2)$$

The channel capacity is achieved by zero-mean complex Gaussian inputs whose covariance matrix maximizes the objective in (4-2). Henceforth in this chapter, we will set $\sigma_v^2 = 1$. To demonstrate how to compute this optimum autocovariance matrix Σ_s^*, let the singular value decomposition representation of H and Σ_s be $H = U_H D_H V_H^\dagger$ and $\Sigma_s = U_s D_s U_s^\dagger$, respectively. Substituting these in (4-2), the objective now becomes

$$\log_2[\det (I + U_H D_H V_H^\dagger \Sigma_s (U_H D_H V_H^\dagger)^\dagger)]$$
$$= \log_2[\det (I + D_H V_H^\dagger U_s D_s U_s^\dagger V_H D_H)]. \quad (4\text{-}3)$$

It is known that choosing $U_s = V_H$ maximizes (4-3) [2]. Substituting $U_s = V_H$ into (4-2), the capacity now becomes

$$C = \max_{tr(\Sigma_s) \le \rho\sigma_v^2} \log_2[\det (I + |D_H|^2 D_s)]$$

$$= \max_{tr(\Sigma_s) \le \rho\sigma_v^2} \sum_{k=1}^n \log_2(1 + |D_H(k, k)|^2 D_s(k, k)). \quad (4\text{-}4)$$

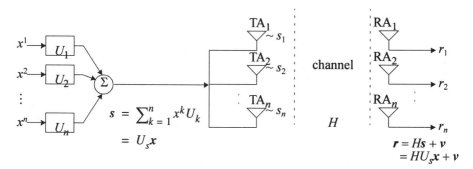

Fig. 4-1. Diagram illustrating the discrete-time input-output relationship of a dual antenna-array system. The transmitted signal s is a linear combination of n orthonormal n-tuples u_k. TA$_i$ stands for the transmitting antenna i and RA$_j$ stands for the receiving antenna j.

The matrix D_s^* that maximizes the summation term in (4-4) is the classical water-filling solution [8]:

$$D_s(k, k)^* = \left(\mu - \frac{\sigma_v^2}{|D_H(k, k)|^2}\right)^+, \qquad (4\text{-}5)$$

were $(g)^+$ denotes the larger of 0 and g, and μ is chosen such that $\sum_{k=1}^{n} D_s(k, k)^* = \rho$. Therefore, the optimum power allocation is $\Sigma_s^* = V_H D_s^* V_H^\dagger$.

The transmitted signal s can be visualized as being obtained by filtering a complex Gaussian n-tuple x with independent entries, as is illustrated in Fig. 4-1. The kth component of x, x^k, has variance $D_s(k, k)$. In this chapter, the term power allocation refers not only to the choice of D_s – the distribution of power among the components of x – but also to the choice of the unitary transformation U_s that transforms x into s. The columns of U_s constitute an orthonormal basis in an n-dimensional space; thus we will refer to U_s as the "transmit basis". For convenience, we will refer to the transmit basis of the optimum power allocation as the "optimum transmit basis".

When the transmitter employs $U_s = V_H$, it leads to a significant advantage, i.e., the receiver can utilize one-dimensional signal processing without loss in capacity [2]. To see this, refer to Fig. 4-1. Let x be an $n \times 1$ vector whose components are independent complex Gaussian random variables with variance $D_s(k, k)$. The vector x is filtered by the unitary transformation U_s to generate the transmitted vector s. The receiver, which also knows the channel, can filter the received signal r by a unitary transformation $U_H^\dagger$. Substituting $s = U_s x$, $y = U_H^\dagger r$, and $v' = U_H^\dagger v$ into (2-2), we obtain a set of n single-input, single-output linear subchannels with i.i.d. $\tilde{N}(0, 1)$ AWGN:

$$y_k = D_H(k, k)x_k + v'_k, \, k = 1, 2, ..., n. \tag{4-6}$$

Clearly, maximum-likelihood estimation of x_k from y_k involves only one spatial dimension.

4.2.2 Uniform Power-Allocation Strategy

In the uniform power allocation, the covariance matrix of s_τ is chosen to be $\Sigma_s = (\rho/n)I_n$. This can done by choosing $D_s = (\rho/n)I_n$. There is no constraint placed on the selection of transmit basis U_s; all unitary U_s lead to $\Sigma_s = U_s D_s U_s^\dagger = (\rho/n)I$. Given the constraint that $\Sigma_s = (\rho/n)I$, the capacity can be obtained by substituting Σ_s into (4-1),

$$C = \sum_{k=1}^n \log_2\left(1 + \frac{\rho}{n}|D_H(k, k)|^2\right). \tag{4-7}$$

Because the capacity does not depend on the transmit basis U_s, uniform power allocation can be used in applications in which the transmitter does not know H.

With uniform power allocation, because the n components of s are statistically independent with the same power, one can treat these n components symmetrically without having to apply different coding, modulation, and signal processing techniques on these n components. Thus it may be desirable from a practical point of view to distribute power uniformly, even when CSI is available at the transmitter. In this case, though the choice of transmit basis does not affect capacity, the optimum transmit basis can still be used to take advantage of the capacity-lossless decomposition of the MIMO channel into SISO subchannels.

4.2.3 Effects of Fading Correlation

The spatial fading correlation cov(H) determines the distribution of the singular values of H which, in turn, determines the distribution of the capacity specified by (4-4) and (4-7). We have shown in Chapter 3 that deriving a closed-form expression for the distribution of the singular values of H is at best tedious and is usually impossible. Here, we study the effects of cov(H) on capacity using a capacity lower bound for the capacity. This lower bound is reasonably close to the exact capacity and leads to useful insights.

Let us review briefly the results on spatial fading correlation. According to the "one-ring" model, in the downlink direction the covariance matrix of H can be well approximated by $cov(vec(H)) \approx \Psi^R \otimes I_n$, where $\otimes$ stands for matrix Kronecker product. The constant matrix Ψ^R can be computed from the physical parameters of the antennas (e.g., antenna spacing), and of the multi-path environments (e.g., angle spread). The distribution of the channel matrix H is therefore well approximated by the distribution of $H_w B^\dagger$ and $B' H_w$ for the downlink and uplink, respectively, where $BB^\dagger = (\Psi^R)'$.

Expressing the channel H as a "colored" version of a white channel H_w enables us to make use of the following properties of matrices with i.i.d. circularly symmetric complex Gaussian matrices:

- For any $n \times n$ unitary matrices U and V, the distribution of $U H_w V^\dagger$ is identical to that of H_w [2].

- Let the QR decomposition of H_w be $H_w = Q_w R_w$. The squared amplitude of the kth element on the diagonal of R_w, $|R_w(k, k)|^2$, is chi-squared distributed with $2(n - k + 1)$ degrees of freedom, and that all the diagonal and upper-diagonal entries of R_w are mutually independent [29]. Note that due to the fact that the real and imaginary parts of any entry of H_w are independently Gaussian distributed with variance $1/2$, the expected value of a chi-squared random variable with $2l$ degrees of freedom is l.

Consider a downlink communication, in which $H \sim H_w B^\dagger$. In (4-1), we substitute for H by $H_w B^\dagger$, and use the singular value decomposition $B^\dagger = U_B D_B V_B^\dagger$ and the QR decomposition $H_w = Q_w R_w$. We obtain

$$C \sim \log_2[\det (I + H_w B^\dagger \Sigma_s B H_w{}^\dagger)]$$
$$\sim \log_2[\det (I + R_w D_B V_B{}^\dagger \Sigma_s V_B D_B R_w{}^\dagger)]. \tag{4-8}$$

By arbitrarily choosing $\Sigma_s = V_B D_s V_B{}^\dagger$, we can formulate a lower-bound of (4-8):

$$C \sim \log_2[\det (I + R_w D_B D_s D_B R_w{}^\dagger)]$$
$$\geq \sum_{k=1}^{n} \log_2(1 + D_s(k, k) D_B^2(k, k) |R_w(k, k)|^2). \tag{4-9}$$

Equation (4-9) indicates that the capacity with optimum power allocation is lower-bounded by the maximal combined capacity of the following system:

n SISO subchannels $y_k = D_B^2(k, k) |R_w(k, k)|^2 x_k + v'_k$, $k = 1, 2, ..., n$, where

- $|R_w(k, k)|^2$ is chi-squared distributed with $2(n - k + 1)$ degrees of freedom,

- x_k is zero-mean circularly symmetric complex Gaussian with variance $D_s(k, k)$,

- v_k' is zero-mean circularly symmetric complex Gaussian with variance σ_v^2, and

- all random variables mentioned ($|R_w(k, k)|^2$, x_k, and v_k') are mutually independent.

Given D_B and R_w, the capacity of this system is again maximized by a water-filling solution D_s.

It is clear that the effect of fading correlation on the capacity lower bound is to scale the power gain of the kth subchannel by $D_B^2(k, k)$. When the fades are i.i.d., $D_B^2(k, k) = 1$ for all k and these n subchannels are scaled equally. When the fades are correlated, the gains of some channels are enhanced while those of others are reduced, because the trace of D_B^2 is equal to n. In the uplink, this interpretation no longer applies, as will be shown in Chapter 4.4. Nevertheless, because the singular values of H and H' are the same, the net effect of fading correlation on capacity when optimum or uniform power-allocation strategy is employed is exactly the same as that in the downlink. The effects of fading correlation on the distribution of the ordered eigenvalues of

$HH^\dagger$ is observed to enhance the first few largest eigenvalues and decrease the rest as well (Chapter 3.4).

4.2.4 Asymptotical Behavior of Channel Capacity with Optimal Power Allocation

We have shown in Chapter 2 that, with uniform power allocation, under the independent fading assumption, the ratio of capacity to the number of antennas of an (n, n) channel converges almost surely to a nonzero constant. Here we will show that with optimum power allocation the ratio of capacity to the number of antennas also converges almost surely to a nonzero number, which is a function of SNR. This result is due to Chuah, Kahn, and Tse [9].

Consider a given n. We randomly generate an instance of an $n \times n$ channel H_n whose entries are i.i.d. $\tilde{N}(0, 1)$. Let F_n be the empirical distribution of the eigenvalues of $H_n H_n^\dagger$. That is, $F_n(\lambda)$ is defined as the fraction of the eigenvalues of $H_n H_n^\dagger$ that is less than or equal to λ. Note that F_n is a function of H_n and is a random function. The following theorem describes the asymptotical property of F_n as $n \to \infty$.

Theorem. Define $G_n(\lambda) \equiv F_n(n\lambda)$. Then almost surely, G_n converges to a nonrandom distribution G^*, which has a density given by

$$g(\lambda) = \begin{cases} \dfrac{1}{\pi}\sqrt{\dfrac{1}{\lambda} - \dfrac{1}{4}}, & 0 \le \lambda \le 4, \\ 0 & \text{otherwise.} \end{cases} \qquad (4\text{-}10)$$

We have already made use of this theorem in Chapter 2. Using this theorem, the asymptotic performance of channel capacity with full CSI at the transmitter can be derived as well.

Theorem. Denote the channel capacity with full CSI at the transmitter given channel H by $C_{\text{opt}}(H)$. Almost surely,

$$\frac{C_{\text{opt}}(H_n)}{n} \to C^* \qquad (4\text{-}11)$$

where

$$C^* = \int_0^4 (\ln(\mu\lambda))^+ \cdot \frac{1}{\pi}\sqrt{\frac{1}{\lambda} - \frac{1}{4}}\,d\lambda \qquad (4\text{-}12)$$

and μ satisfies

$$\int_0^4 \left(\mu - \frac{1}{\lambda}\right)^+ \cdot \frac{1}{\pi}\sqrt{\frac{1}{\lambda} - \frac{1}{4}}\,d\lambda = \rho. \qquad (4\text{-}13)$$

The results in (4-12) and (2-7) can be combined to obtain the benefit of optimal power allocation over uniform power allocation when $n \to \infty$ [9]. When $n \to \infty$, $C_{opt}(H_n)/n \to C^*$ and $C_{uni}/n \to C_{uni}^*$. Using L'Hopital's rule, it can be shown that at low SNR,

$$\lim_{\rho \to 0} \frac{C^*}{C_{uni}^*} \to 4. \qquad (4\text{-}14)$$

At high SNR,

$$\lim_{\rho \to \infty} \frac{C^*}{C_{uni}^*} \to 1. \qquad (4\text{-}15)$$

4.3 Blind Transmission Systems

A blind transmission system cannot employ optimum power-allocation strategy due to the lack of instantaneous CSI. Although uniform power-allocation strategy can be used to achieve robust performance against channel uncertainty, if the spatial fading correlation is high it results in a significant loss in capacity when compared to using optimum power-allocation strategy. Conceptually, the capacity loss arises mainly because, with a uniform power distribution, part of the transmitted power is allocated to subchannels with low gains. The goal of this section is to devise a nonuniform power allocation, which we refer to as stochastic water-filling power allocation, that avoids this inefficient use of power.

The key assumption here is that while transmission is blind, the transmitter does know the spatial fading covariance cov(H). Because spatial fading statistics is locally stationary, and hence varies much more slowly than the channel itself, in many applications it is realistic for the transmitter to acquire this knowledge. In the following, the downlink and uplink scenarios are studied separately. Although the direction of transmission does not affect the performance and applicability of both optimum and uniform power-allocation strategies, we will show that, due to the structure of the spatial fading covariance, the stochastic water-filling power allocation achieves a higher average capacity than the uniform power allocation only in the downlink.

4.3.1 Stochastic Water-Filling in the Downlink

We study this power allocation problem using the capacity lower bound. Let the power allocation be $\Sigma_s = U_s D_s U_s^\dagger$. Note that neither U_s nor D_s can be a function of H. Let the QR decomposition of the matrix product $B^\dagger U_s$ be $B^\dagger U_s = Q_{BU} R_{BU}$. Similar to (4-8), the capacity is lower-bounded by:

$$C \sim \log_2[\det{(I + H_w B^\dagger \Sigma_s B H_w^\dagger)}]$$
$$\geq \sum_{k=1}^{n} \log_2(1 + D_s(k, k) |R_{BU}(k, k)|^2 |R_w(k, k)|^2). \tag{4-16}$$

The RHS of the inequality (4-16) is the sum of the capacities of n SISO sub-channels with power gains $|R_{BU}(k, k)|^2 |R_w(k, k)|^2$, $1 \leq k \leq n$. Comparing equations (4-8) and (4-16), when the CSI is not available to the transmitter, there is a penalty due to the mismatch of the transmit basis with the channel. Because R_w is unknown to the transmitter, instead of directly maximizing the RHS of (4-16), the transmitter chooses U_s and D_s to optimize a chosen statistical property of the RHS of (4-16). The optimal choice of power allocation without CSI is still an open question. In the following, we describe a two-step approach to determine a good choice of power allocation (U_s, D_s).

The transmitter first chooses the transmit basis U_s. Note that because $D_s(k, k)$ is decreasing in k and $|R_w(k, k)|^2$ is chi-squared distributed with degrees of freedom that are decreasing in k, to achieve the highest average capacity one must choose U_s such that $|R_{BU}(k, k)|^2$ is also decreasing in k. Consider choosing U_s according to the following greedy procedure. The set of unitary matrices that yield the highest value of $|R_{BU}(1, 1)|$ is first identified. We then find the largest subset of this set whose members yield the largest value of $|R_{BU}(2, 2)|$, and so on. The procedure ends when we finally obtain the set whose member yield the largest value of $|R_{BU}(n, n)|$. It can be easily shown that V_B belongs to this set. If $U_s = V_B$ is used, $|R_{BU}|^2 = D_B^2$.

In principle, once U_s is selected, we can calculate the corresponding D_s that maximizes some chosen statistical property of channel capacity. A reasonable choice is to maximize the expectation of the upper-bound of the right-hand side of equation (4-16). Recall that $|R_w(k, k)|^2$ is chi-squared distributed with $2(n - k + 1)$ degrees of freedom, so that $E[|R_w(k, k)|^2] = n - k + 1$. Furthermore, because $E(\log(X)) \leq \log(E(X))$,

$$E\left\{\sum_{k=1}^{n} \log_2(1 + D_s(k, k)|R_{BU}(k, k)|^2 |R_w(k, k)|^2)\right\}$$

$$\leq \sum_{k=1}^{n} \log_2(1 + D_s(k, k)|R_{BU}(k, k)|^2 E[|R_w(k, k)|^2]) \qquad (4\text{-}17)$$

$$= \sum_{k=1}^{n} \log_2(1 + D_s(k, k)|R_{BU}(k, k)|^2 (n - k + 1)).$$

The power allocation that maximizes the term to the right of equality in (4-17) is again solved using the water-filling procedure. That is,

$$D_s^*(k, k) = \left(\mu - \left(\frac{1/|R_{BU}(k, k)|^2}{n - k + 1}\right)\right)^+ \text{ where } \sum_{k=1}^{n} D_s^*(k, k) = \rho. \quad (4\text{-}18)$$

In summary, the transmitter chooses $U_s = V_B$ according to the greedy criterion, and it distributes power according to (4-18). We will refer to such a practice as stochastic water-filling. The stochastic water-filling power allocation can provide capacity improvement in highly correlated fading environments. We will compare the capacity achieved by stochastic water-filling power allocation with those achieved by optimum and uniform power allocation strategies in Chapter 4.5.

4.3.2 Optimality of Uniform Power Allocation in the Uplink

Unlike the downlink direction, uniform power allocation achieves the highest average capacity in the uplink direction.

Lemma. For any positive semi-definite diagonal matrix D_B and D_s, the distribution of singular values of the matrix $D_B H_w D_s$ is identical to that of the matrix $D_B H_w M D_s M$, where M is any square permutation matrix, i.e. exactly one entry in each column and each row is equal to one and the other entries are zero.

Proof. The lemma is evident because that the distribution of H_w is identical to the distribution of $H_w M$ and that the singular values of X and XM are identical for any matrix X.

To show the optimality of uniform power allocation, we substitute $H \sim B' H_w$ for H in (4-1):

$$C(\Sigma_s) \sim \log_2[\det (I + B' H_w \Sigma_s H_w^\dagger (B')^\dagger)]$$
$$\sim \log_2[\det (I + D_B H_w D_s H_w^\dagger D_B)], \tag{4-19}$$

where $tr(D_s) \le \rho$. The goal here is to prove that $D_s{}^* = (\rho/n)I$ achieves the highest expected value of the RHS of (4-19). Consider a particular choice of nonnegative diagonal matrix D_s. Let $D_s^{(l)}, l = 1, 2, \ldots, n!$, denote the $n!$ possibly distinct diagonal matrices whose diagonal entries are permutations of the diagonal entries of D_s. According to the lemma, using any one of the $D_s^{(l)}$ in (4-19) leads to the same capacity distribution. The mean value of capacity using D_s is upper-bounded by

$$E(C(D_s)) = E\left(\frac{1}{n!}\sum_{l=1}^{n!} C(D_s^{(l)})\right)$$
$$\le E\left(C\left(\frac{1}{n!}\sum_{l=1}^{n!} D_s^{(l)}\right)\right) = E\left(C\left(\frac{\rho}{n} I_n\right)\right). \tag{4-20}$$

The inequality in (4-20) results from the concavity of the logarithm function.

4.4 Capacity Penalty From One-dimensional Processing of Multi-dimensional Signals

In blind transmission systems, because the optimum transmit basis is not known by the transmitter, the capacity-lossless decomposition of the MIMO channel into n SISO subchannels outlined in Chapter 4.2.2 is not possible.

Nevertheless, it is still desirable to extract individual observations of the components of x from the received signal, thereby one-dimensionalizing the inherently multi-dimensional signal-processing problem, as long as the penalty in capacity is not too high. This is essentially a trade-off between receiver complexity and required SNR. The study of this problem was pioneered by Foschini, who showed that the capacity penalty per spatial dimension associated with the layered space-time (LST) architecture approaches zero as n approaches infinity in an independent fading environment [26].

In this section, we focus on blind transmission systems that employ one-dimensionalization techniques. The main goal is to investigate the capacity penalty associated with one-dimensionalization in systems that use stochastic water-filling and uniform power-allocation strategies. As will be pointed out later in this section, the study of one-dimensionalization techniques gives operational explanations to the capacity lower bounds (4-9) and (4-16).

4.4.1 ZF and MMSE Successive Interference Cancellation

We again start with the relation between x and r shown in Fig. 4-1:

$$r_\tau = HU_s x_\tau + v_\tau. \qquad (4\text{-}21)$$

In the following, we omit the subscript τ for simplicity. To derive individual observations of the components of x, the simplest scheme is to multiply r by the inverse[1] of HU_s:

$$r' = (HU_s)^{-1} r = x + (HU_s)^{-1} v . \qquad (4\text{-}22)$$

Clearly, the kth component of r' – the observation of x_k – is x_k corrupted by additive noise. Because the observation of x_k contains no interference from other components of x, this is referred to a zero-forcing (ZF) channel inversion. Directly inverting the channel, as in (4-22), can lead to deleterious noise enhancement. To alleviate noise enhancement, one can use successive interference cancellation (SIC) in conjunction with channel inversion. In SIC, the receiver extracts the observations of $\{x^1, x^2, ..., x^n\}$ one-by-one, following some specific order. After obtaining the observation of x^k, the receiver makes

---

1. When $m > n$, use the left pseudo-inverse of HU_s instead of the inverse.

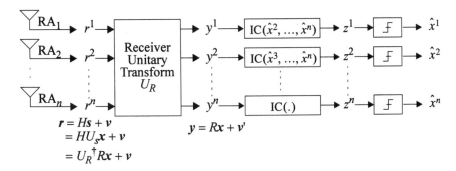

Fig. 4-2. Diagram illustrating ZF SIC. IC stands for interference cancellation.

a decision on x^k and cancels out the contributions of x^k in r. We assume here that the observations of x^k are extracted in order of descending k. A mathematical representation of combined ZF channel inversion and SIC is as follows. We denote the QR decomposition of HU_s by $HU_s = U_R^\dagger R$. The receiver first passes the received signal r through U_R to obtain an n-tuple y,

$$y = U_R r = Rx + v', \qquad (4\text{-}23)$$

where $v' = U_R v$ is an n-tuple whose distribution is identical to that of v. Note that because R is upper triangular, excluding the noise component, y^k is a linear combination of $x^k, x^{k+1}, ..., x^n$. The contributions of $x^{k+1}, ..., x^n$ in y_k are canceled using SIC. Assuming that there are no errors in the cancellation process, the individual observations of x^k are:

$$z^k = R(k, k)x^k + v'(k), \; k = 1, 2, ..., n. \qquad (4\text{-}24)$$

Let the variance of x^k be ρ^k. The signal to noise ratio of z^k is

$$snr_k = |R(k, k)|^2 \rho_k. \qquad (4\text{-}25)$$

The block diagram of ZF SIC is shown in Fig. 4-2. Although we have chosen to detect x^k in order of descending k, by re-ordering the x^k prior to detection, our analysis is applicable to detection in any order. A prototype dual antenna-array system utilizing ZF SIC has been demonstrated [5].

We can also use other channel inversion criteria, such as the minimum mean-squared error (MMSE) criterion, in conjunction with SIC. To obtain the linear MMSE estimate of x^n from r, one partitions HU_s as $HU_s = \left[H_{n-1} \mid h_n \right]$. Equation (4-22) can be written as

$$r = \left[H_{n-1} \mid h_n \right] \begin{pmatrix} x_1^{n-1} \\ x_n \end{pmatrix} + v. \tag{4-26}$$

Let $\Sigma_H = (HU_s)\Sigma_x(HU_s)^\dagger$. The linear least-square estimate of x^n is

$$z^n = \rho_n h_n^\dagger (\Sigma_H + I_n)^{-1} r = \{ \rho_n h_n^\dagger (\Sigma_H + I_n)^{-1} h_n \} x^n + \text{error} , \tag{4-27}$$

and the error variance in z^n is

$$\rho_n \left(1 - h_n^\dagger \left(\frac{\Sigma_H + I_n}{\rho_n} \right)^{-1} h_n \right) . \tag{4-28}$$

Let $\hat{x}^n$ denote the decision made on x^n. We then replace HU_s by H_{n-1}, x by x_1^{n-1}, and r by $r - \hat{x}^n h_n$ in (4-26) to (4-28) to obtain the linear least square estimate of x^{n-1}. The process is repeated until all n elements of x have been observed.

In the subsequent analysis, we assume that the receiver uses ZF SIC.

4.4.2 Downlink Analysis

The distribution of capacity is determined by the distributions of snr_k, $1 \le k \le n$, defined in equation (4-25). Similar to the derivations in Chapter 4.3.1, in the downlink direction, the distribution of HU_s can be easily shown to be identical to that of $Q_w R_w R_{BU}$ by substituting $H_w B^\dagger$ for H and the QR decompositions of $B^\dagger U_s$ and H_w:

$$HU_s \sim H_w B^\dagger U_s \sim H_w Q_{BU} R_{BU} \sim H_w R_{BU} \sim Q_w R_w R_{BU} .$$

Substituting $HU_s = Q_w R_w R_{BU}$ into (4-23), noting that $U_R = Q_w^\dagger$, we have:

$$y = Q_w^\dagger r = R_w R_{BU} x + v' .\tag{4-29}$$

The interference-free observation of x^k is

$$z^k = R_w(k, k) R_{BU}(k, k) x^k + v'(k) .\tag{4-30}$$

The downlink capacity with ZF SIC is thus distributed as

$$C \sim \sum_{k=1}^{n} \log_2(1 + \rho_k |R_{BU}(k, k)|^2 |R_w(k, k)|^2) .\tag{4-31}$$

By comparing (4-31) with (4-16), we can explain the capacity lower bound in (4-16) as the capacity achieved with a suboptimal processing architecture (ZF SIC) at the receiver. The stochastic water-filling power allocation derived in (4-18) can be employed here to achieve a capacity higher than uniform power allocation.

4.4.3 Uplink Analysis

Similar to the above analysis, we substitute H by $B'H_w$ in (4-21) to obtain the distributions of snr_k for the uplink. Because U_s does not affect the distribution of $HU_s \sim B'H_w U_s = B'H_w$, U_s can be simply chosen as the identity matrix. Equation (4-21) becomes

$$r = B'H_w x + v .\tag{4-32}$$

To highlight the differences between the distributions of snr_k in the uplink and downlink, we will use a different mathematical representation of SIC. Substituting $H_w = Q_w R_w$ into (4-32) and then premultiplying by $(B'Q_w)^{-1}$ on both sides of (4-32), we get

$$y = (B'Q_w)^{-1} r = R_w x + u ,\tag{4-33}$$

where u is an additive white Gaussian noise vector with covariance $\Sigma_u = Q_w^\dagger (B^* B')^{-1} Q_w$. The autocovariance matrix Σ_u is random and is independent of R_w. Because Σ_u is, in general, not diagonal, the components of noise u are, in general, not independent. Equation (4-33) can be written component-wise as

$$y^k = R_w(k, k)x^k + \{\text{interference from } x^{k+1}, ..., x^n\} + u^k,$$

$$k = 1, 2, ..., n. \tag{4-34}$$

When the receiver makes the decision on x^k, it also obtains an estimate on the additive noise u^k: $\hat{u}^k = y^k - \sum_{l=k}^{n} R_w(k, l)\hat{x}^l$. Assuming that the noise decisions are all correct, i.e., $\hat{u}^l = u^l$ for $l = k, ..., n$, these decisions can be used to form a minimum mean square estimate of u^{k-1}, which is simply the conditional expectation of u^{k-1} given $\hat{u}^k, ..., \hat{u}^n$ because $u^{k-1}, ..., u^n$ are jointly Gaussian. In other words, in this formulation, besides canceling the interference term in (4-34), SIC also removes the predicted noise component. The interference-free observation of x_k provided by SIC is thus

$$z^k = R_w(k, k)x^k + (u^k - E(u^k|\hat{u}^{k+1}, ..., \hat{u}^n)), k = 1, 2, ..., n. \tag{4-35}$$

The signal power in z^k is $|R_w(k, k)|^2 \rho_k$. To obtain the noise variance, we partition Σ_u:

$$\Sigma_u = \begin{bmatrix} \cdots & & \cdots \\ & \Sigma_u(k, k) & a_k \\ \cdots & & \\ & a_k^\dagger & \Sigma_u^{k+1} \end{bmatrix}. \tag{4-36}$$

The variance of noise in z_k is

$$\sigma_k^2 = \Sigma_u(k, k) - a_k(\Sigma_u^{k+1})^{-1}a_k^\dagger, \tag{4-37}$$

which is a function of Q_w and thus is itself a random variable.

The difference in the uplink and downlink is apparent when (4-30) and (4-35) are compared. The distributions of the signal power gain, $|R_w(k, k)|^2$, are the same. In the downlink, the noise power $E[|v'(k)|^2]$ is a constant. In contrast, in the uplink, the noise power as formulated in (4-35) is random and can be much larger than its counterpart in the downlink. Furthermore, the variance of the noise power can be much higher than the variance of the signal power gain $|R_w(k, k)|^2$. (Note that σ_k^2 is independent of R_w.)

For example, under the correlated fading profile used in Chapter 4.5, with $n = 16$, in the downlink the standard deviation of snr_k is the standard deviation of $|R_w(k, k)|^2$, which is 1.26 dB for $k = 5$. In the uplink, the standard deviation of snr_k is the standard deviation of $|R_w(k, k)|^2 / \sigma_k^2$, which is 7.2 dB for $k = 5$. The high variability of snr_k have a profound implication on the system design. The channel codes applied on $\{x_\tau^k\}$ must be designed to sustain a good performance even when the channel realization is adverse, which has a much higher probability in the uplink due to the higher variance of snr_k. One solution is to employ channel codes on the sequences $\{x_\tau^{(\tau \bmod n) + 1}\}$, rather than on $\{x_\tau^k\}$ [26].

4.5 Capacity Results

In this section, we present the channel capacities and their respective lower bounds for two extreme fading correlation situations. In the first, that of independent fading, all the components of H are i.i.d. In the second, that of strongly correlated fading, we use the following parameters in the "one-ring" model. We assume that linear antenna arrays are used, with a transmitting antenna spacing of two wavelengths. The angle of arrival is 0°, and the angle spread is 0.6°.

For each configuration, 10,000 independent random channel realizations are generated to obtain the histograms of the investigated quantity. Specifically, for the correlated fading scenario, we first obtain B using the scatterer model parameters specified above. The channel samples are then generated by multiplying independent, randomly generated H_w by $B^\dagger$. We use the Monte-Carlo approach here because closed-from expressions for the distributions of channel capacity are very difficult to derive.

An important performance measure for a dual antenna-array system operating in this burst mode is the capacity at a given outage probability q, denoted as C_q. In other words, the probability that the channel capacity of a randomly chosen H is lower than C_q is q. In this chapter, comparisons among different schemes will be presented, when possible, based on the capacity at ten-percent outage, $C_{0.1}$. We will sometimes use the average capacity as a criterion for comparison.

In Fig. 4-3, we plot $C_{0.1}$ vs. n at SNR $= \rho = 18$ dB, while in Fig. 4-4, we plot $C_{0.1}$ vs. SNR for $n = 16$. From these two figures, we can observe the following:

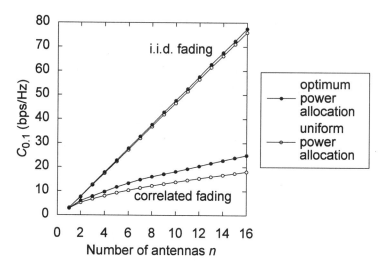

Fig. 4-3. Capacity of dual antenna-array systems as a function of *n*. Here, SNR is 18 dB. (a) The following parameters are used in the "one-ring" model for the fading covariance: transmitting antenna spacing is two wavelengths long; the angle of arrival is 0° (broadside); and the angle spread is 0.6°.

1. It has been established that asymptotically the average capacity per spatial dimension converge to a constant depending only on SNR [9], [26]. From Fig. 4-3, we see that with independent fading, the relationship between $C_{0.1}$ and *n* is approximately linear even when *n* is small.

2. Fading correlation can significantly reduce $C_{0.1}$. This can be explained using (4-9) by noting that in this setting when $n \geq 4$ the largest singular value of D_B is more than 20dB higher than the fourth largest one. Thus in the capacity lower bound (4-9), the transmit power allocated to the *k*th sub-channel for $k \geq 4$ is not effective in conveying information.

3. The difference between capacities with optimum power allocation and with uniform power allocation is significant only when the fading correlation is high. This implies that the additional complexity of optimum power allocation over uniform power allocation is justified only if the fades are strongly correlated.

In Fig. 4-5, we compare channel capacities and their corresponding lower bounds computed using (4-9). The lower bound is seen to be reasonably close to the capacity throughout the SNR range that we consider, in both independent and correlated fading environments. In particular, when the fades are highly correlated, with uniform power allocation the $C_{0.1}$ curve and its lower bound are very close.

In Fig. 4-6, we also plot the capacity lower bounds (the dashed curves). They represent the capacities achieved with a ZF SIC receiver using uniform and stochastic water-filling power allocation assuming that $U_s = V_B$ is chosen. The gap between the dashed curve and the corresponding solid curve indicates the capacity penalty due to one-dimensionalization. The penalty is not significant for the range of parameters that we consider.

4.6 Summary

In this chapter, we studied three power-allocation strategies for dual antenna-array systems, using primarily the information-theoretic capacity as

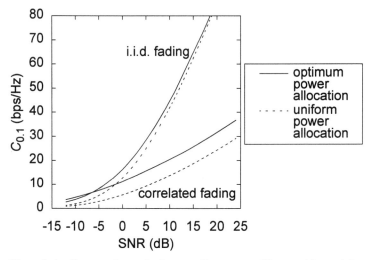

Fig. 4-4. Comparison between the capacities achieved by optimum power allocation and uniform power allocation when $n = 16$.

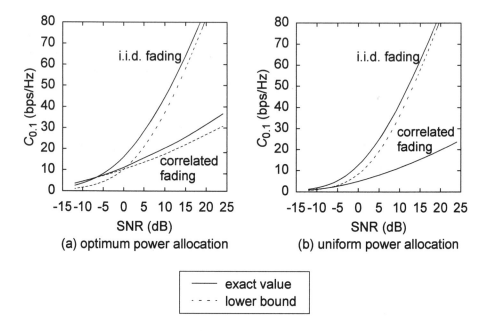

Fig. 4-5. Capacity of dual antenna-array systems using optimum and uniform power allocations, and their respective lower bounds.

the performance criterion. Specifically, we focused on the performance of dual antenna-array systems in environments where the fades are highly correlated. In general, with medium-to-high SNR, the higher the fading correlation, the lower the capacity.

When the transmitter knows the instantaneous channel realization, optimum power allocation, which achieves the highest capacity throughput that the particular channel realization supports, can be used. Optimum power allocation refers to the use of a particular transmit basis and a power distribution that is computed using the water-filling algorithm.

Although uniform power allocation does not achieve a capacity as high as optimum power allocation, it offers many practical advantages. One major advantage is that the capacity is not dependent on the choice of transmit basis, making uniform power allocation suitable for blind transmission systems. Furthermore, when the fades are i.i.d., the difference between the capacities

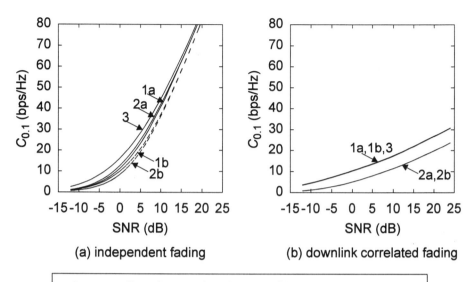

(a) independent fading   (b) downlink correlated fading

1a: capacity using stochastic water-filling power allocation
1b: lower bound of 1a
2a: capacity using uniform power allocation
2b: lower bound of 2a
3: capacity using optimum power allocation

Fig. 4-6. Capacity achieved by using stochastic water-filling power allocation strategy. The capacity achieved using uniform and optimum power allocation strategies are plotted for comparison purposes. The lower bound curves 1b and 2b also represent the channel capacities achieved when ZF SIC is employed in the receiver with the corresponding power allocation strategies.

achieved by optimum and uniform power allocations is small. However, when the fades are correlated, the difference can be large.

We have demonstrated a nonuniform power allocation that achieves a capacity close to that achieved by optimum power allocation in the downlink when the fading correlation is high. This power allocation is calculated via the stochastic water-filling procedure, which requires the transmitter to know only the fading statistics, not the instantaneous channel realization. Our conclusion is that on the downlink a blind transmission system can employ uniform power allocation when fading correlation is low and stochastic waterfilling power allocation when fading correlation is high, and can thereby obtain performance close to the optimum power allocation. On the uplink, however, uniform power allocation achieves the highest average capacity for blind transmission systems.

We also extended our analysis to systems that employ techniques such as SIC to reduce the receiver complexity. The key idea is to one-dimensionalize the inherently multi-dimensional signal processing task. Our result shows that the capacity penalty incurred by using ZF SIC is small over the range of physical parameters and power-allocation strategies that we considered.

Layered Space-Time Codes: Analysis and Design Criteria

5.1 Introduction

In the previous two chapters, we focus on unveiling the potential of dual antenna-array systems. We have shown that asymptotically if the fades are independent and if $n \leq m$, the average channel capacity of an (n, m) channel is $O(n)$.

In practice, channel codes are necessary to provide a throughput that is close to capacity with a reasonable error probability. Because the transmitter has multiple transmit antennas, the channel codes to be employed also has multiple spatial dimension. Thus, these channel codes are referred to as space-time codes. In this chapter, we are interested in space-time codes whose throughput is proportional to n, assuming that $n \leq m$. In other words, when such a space-time code is used in an (n, m) system, the bit-rate of the system scales linearly in n. This assumption on bit rate will be made implicitly throughout this chapter. We will not consider the class of space-time codes that maintain a throughput independent of n; e.g., [31] - [33], and the smart-greedy codes [34]. The goal of these codes is to leverage the transmit diversity to reduce the required SNR.

An important consideration for space-time codes is the decoding complexity. In many applications, the number of antennas can indeed be very large. Assume, for simplicity, that $n = m$. Consider a family of nontrivial space-time codes whose throughput is proportional to n. As will be shown in Chapter 5.2, the complexity of decoding such a space-time code according to a maximum-likelihood (ML) criterion is generally exponential in n. When n is large, space-time codes that admit low complexity (suboptimal) decoding algorithm are very desirable.

In this chapter, we propose a class of space-time codes whose throughput and decoding complexity scale linearly and quadratically with n, respectively. Such a space-time code is constructed based on the layered space-time (LST) architecture proposed by Foschini in [26]; therefore, we refer to this class of space-time code as an LST code. There are two types of LST architectures: horizontally layered space-time architecture (HLST) and diagonally layered space-time architecture (DLST). In addition to low decoding complexity, LST codes offer the advantage of utilizing the established 1-D codec technology. The use of a suboptimal decoding scheme, however, does incur a power penalty compared to ML decoding. To date, studies on LST codes have focused on the information-theoretic considerations. Another class of codes that has been proposed as low-complexity space-time codes is the class of codes that admit multistage decoding [34] [35].

In this chapter, we first analyze the error performance of LST codes. We consider both slow and fast fading environments, as well as both high and low SNR regimes. We derive the key parameters that dominate the error performance, and propose design criteria for LST codes. From the error analysis, we find that DLST outperforms HLST in slow fading environments. For DLST codes, the optimum trade-off among several design parameters is presented. We also quantify the power penalty incurred by LST decoding compared to ML decoding.

We then examine the operational aspects of DLST codes, specifically the use of block codes and convolutional codes as the constituent codes for DLST codes. With convolutional constituent codes, we will show that the original DLST architecture does not lead to satisfactory performance. We propose the single-stream structure as the solution. For block constituent codes, we show that permuting the order of symbols in a block codeword results in dramatic differences in performance. Therefore, the optimum permuting order should be employed. The error analysis and design criteria for these modified LST codes are provided. These solutions achieve greatly improved performance.

The remainder of this chapter is organized as follows. In Chapter 5.2, the notation of space-time codes is introduced. We also provide the error probability analysis for ML decoding. In Chapter 5.3, we introduce the LST architecture, in particular the HLST and DLST architectures. The decoding complexity of LST codes is shown to be quadratic in n. In Chapter 5.4, we analyze the performance of LST codes. We demonstrate that DLST is superior to HLST, especially in slow fading environments. The design criteria, optimum design choices, and penalties associated with the suboptimal decoding mechanism are presented. In Chapter 5.5, we examine the operational aspects of DLST codes, and propose the modified structures to achieve improved performance. We also present example DLST codes. We give concluding remarks in Chapter 5.6.

5.2 Space-Time Codes

5.2.1 Notation

Consider the encoding process first. The encoder at the transmitter applies the channel code to the input information bits to generate an n-row (possibly semi-infinite) matrix C. The kth row, τth column element of C, denoted by c_τ^k, represents the signal to be transmitted by antenna k at time slot τ. Such a channel code differs from conventional channel codes in that it involves multiple transmit spatial dimensions. To emphasize this distinction, it is referred to as a space-time code.

At the receiver, during the time slot τ, the receiving antenna l receives a signal r_τ^l. This received signal r_τ^l contains a superposition of transmitted signals c_τ^k, $k = 0, 1, ..., n-1$, and an AWGN component v_τ^l. For a narrowband flat-fading channel, the gain connecting transmitting antenna k and receiving antenna l at time τ can be denoted by a complex number $h_\tau^{l \leftarrow k}$. We define the vectors $c_\tau = (c_\tau^1 \ c_\tau^2 \ ... c_\tau^n)'$, $r_\tau = (r_\tau^1 \ r_\tau^2 \ ... r_\tau^m)'$, $v_\tau = (v_\tau^1 \ v_\tau^2 \ ... v_\tau^m)'$. The discrete-time, input-output relation of the (n, m) dual antenna-array system over a narrowband flat-fading channel can be written in the following vector notation:

$$r_\tau = H_\tau c_\tau + v_\tau . \qquad (5\text{-}1)$$

In this chapter, we will always assume that $n \leq m$.

The following terminology is used in this chapter. The matrix C, which is the coded matrix output of the transmitter encoder, is referred to as a space-time *codeword matrix*. A space-time codeword matrix can be thought of as a serial concatenation of *n-tuples*, and an *n*-tuple is composed of n *symbols*. Note that the first row of the matrix C is indexed as row zero, not row one.

To facilitate the comparison between dual antenna-array systems using space-time codes and single transmit-antenna systems using conventional 1-D channel codes at equal average transmit powers (total over all transmit antennas), the average energy of an *n*-tuple is E, regardless of the spatial dimensionality n.

5.2.2 Space-Time Codes with ML Decoding: Performance and Complexity

ML decoding is optimum in terms of achieving the lowest error probability. In the following, expressions for the error probability of a space-time code with ML decoding are given. These will be used as a reference to quantify the performance of other decoding mechanisms. The error probability with ML decoding has been derived independently by Tarokh et. al. [34] [35] and Guey et. al [36]. The special case when $n = 1$ was derived even earlier by Divsalar and Simon [37] [38]. Our matrix notation leads to an elegant derivation; see Appendix I.

Let C and E be two distinct space-time codeword matrices. Suppose that C is the transmitted space-time codeword matrix. The average pairwise error probability between codewords C and E, denoted by $Prob(C \rightarrow E)$, is the average probability that the likelihood of received signal given E is higher than that of the received signal given C. Here, the average is taken over random realizations of H. In a fast fading environment, $Prob(C \rightarrow E)$ is upper-bounded by

$$Prob(C \rightarrow E) \leq \prod_{\tau = 1}^{L} \left(1 + |c_\tau - e_\tau|^2 \frac{E}{4N_0} \right)^{-m}. \qquad (5\text{-}2)$$

In a slow fading environment, $Prob(C \rightarrow E)$ is upper-bounded by

$$Prob(C \rightarrow E) \leq \prod_{k = 1}^{rank(C - E)} \left(1 + \Lambda_k \frac{E}{4N_0} \right)^{-m}, \qquad (5\text{-}3)$$

where Λ_k are the eigenvalues of the matrix $(C - E)(C - E)^\dagger$.

Although ML decoding achieves the lowest error probability, the complexity of implementing ML decoding may be a concern. Consider space-time codes that encode $an + o(n)$ bits per n-tuple, where a is a positive constant. The complexity of ML decoding is generally exponential in n. To see this, consider the log likelihood of receiving y_τ when an n-tuple c_τ is transmitted. The log likelihood is an affine function of $|r_\tau - Hc_\tau|^2$. For a non-trivial H, $|r_\tau - Hc_\tau|^2$ cannot be further reduced; thus an exhaustive search among the $2^{an + o(n)}$ possible choices of c_τ is required. It is mainly the high complexity of conventional ML decoding that motivates the study of space-time codes whose structure allows for efficient ML decoding or low complexity suboptimal decoding that does not incur a significant degradation in error performance.

5.3 Layered Space-Time Architecture

In this section, we present a brief summary of the layered space-time (LST) architecture [26].

5.3.1 Encoding

In the LST architecture, the multi-spatial dimensional signal is obtained by spatially multiplexing 1-D signals, or more generally space-time signals with a spatial dimension l such that l divides n, in a systematic fashion with the goal of reducing the receiver complexity.

The encoding process is illustrated in Fig. 5-1. The first step is to generate the 1-D signals. In Fig. 5-1(a), the input information bit sequence is first demultiplexed into n subsequences, and each subsequence is subsequently encoded by a 1-D encoder. These 1-D channel codes are referred to as the constituent codes. The output of the constituent coder k is a sequence of symbols s_τ^k, $\tau = 0, 1, \ldots$.

The second step is to designate when, and from which antenna, a coded symbol, say s_τ^k, is to be transmitted. One intuitive assignment rule is to always transmit the output coded symbol from constituent encoder k using the transmit antenna k. This is illustrated in Fig. 5-1(b). Under this assignment rule, the space-time codeword has an obvious horizontally layered structure. It

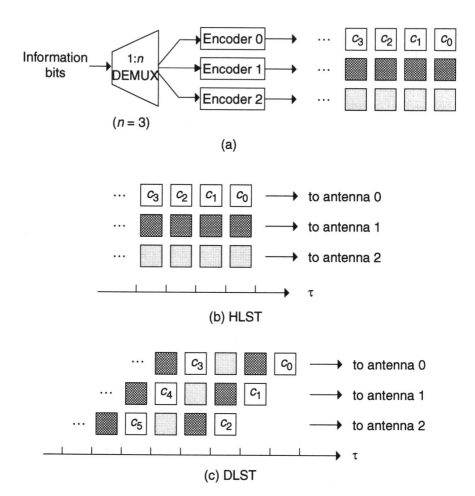

Fig. 5-1. Encoding of LST codes. Here, $n = 3$. Each square represents a symbol. (a) The incoming information bit sequence is first demultiplexed into n subsequences. Each subsequence is encoded using a constituent code. (b) In HLST, the coded symbols from constituent encoder k are transmitted by antenna k. (c) In DLST, the coded symbols from a constituent encoder are transmitted by the n transmitting antennas in turn.

is called the horizontally layered space-time (HLST) architecture. HLST was originally proposed by Foschini in [26]. Another assignment rule, also proposed in [26], is the diagonally layered space-time (DLST) architecture. In DLST, instead of always feeding the output symbols from a constituent coder to a particular transmit antenna, they are fed to the n transmitting antennas in turn. The practice of rotating the roles of antennas is called cycling. DLST is illustrated in Fig. 5-1(c).

If a coded symbol s is to be transmitted at time τ from antenna k, in our notation it is equivalent to assigning the (k, τ)th component of the transmitted space-time codeword matrix C to be s. An informal way of saying this is that the symbol s is used to fill the (k, τ)th slot of C. In HLST, the output of the constituent coder k is used to fill the kth row of the codeword matrix C, i.e., $C_\tau^k = s_\tau^k$. In DLST, the outputs of the constituent coders are used to fill the NW-SE diagonals of C from left to right in turn. Specifically, the output of the constituent coder k fills the $(k + l \cdot n)$th diagonals of C, where $l = 1, 2, \dots$.

5.3.2 Decoding

At the receiver, the received signal is a superposition of transmitted coded symbols scaled by the channel gain and corrupted by AWGN. Instead of decoding the n constituent codes jointly, in the LST architecture, *interference suppression* and *interference cancellation* are employed so that the constituent codes can be decoded individually.

Consider the processing along the spatial dimension first. Let us focus on a given instance in time, say τ. The transmitted n-tuple is c_τ and the received m-tuple is $r_\tau = H_\tau c_\tau + v_\tau$. The goal here is to determine the values of the n components of c_τ, i.e. c_τ^0, c_τ^1, ..., c_τ^{n-1}, with the only available information being r_τ and H_τ. In the LST architecture, the decisions on the values of these n components are made sequentially according to a pre-determined order. Without loss of generality, in this chapter we assume that the decision order is in descending order of the superscript of c_τ^k.

The symbol c_τ^{n-1} is the first one to be decided. To decide the value of the symbol c_τ^{n-1}, a decision variable, denoted by z_τ^{n-1}, is extracted from the received signal r_τ. This decision variable z_τ^{n-1} should contain a low level of interference from other symbols c_τ^{n-2}, c_τ^{n-3}, ..., c_τ^0. Hence, this operation is often referred to as interference suppression. The decision on c_τ^{n-1} is then made. Making use of the decision on c_τ^{n-1}, the receiver can modify the received signal r_τ by removing the contribution of c_τ^{n-1} to it. This modifying operation is referred to as interference cancellation. The process of extracting

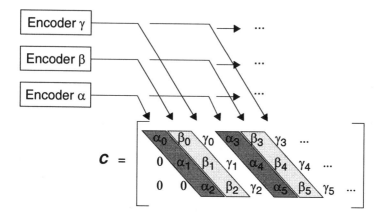

Fig. 5-2. A DLST codeword matrix **C**. Here, $n = 3$.

a decision variable, making a decision on the value, and modifying the received signal is repeated for the remaining symbols c_τ^{n-2}, c_τ^{n-3}, ..., c_τ^0.

The decoding algorithm in the LST architecture utilizes the spatial processing described above and exploits the temporal redundancy of the constituent codes to provide reliable decisions. To decode an HLST code, the receiver first extracts the decision variables for the symbols of the bottommost row of **C**. The resulting decision variable sequence, $\{z_\tau^{n-1}\}$, $\tau = 0, 1, ...$, is used by a conventional 1-D decoder of the corresponding constituent code to produce the decisions on the symbols of this row. The receiver then uses the decision to modify the received signal sequence $\{r_\tau\}$, and then proceeds to decode row $n-2$, $n-3$, and so on. In short, the HLST codeword matrix **C** is decoded row by row, or layer by layer, from bottom to top.

A DLST code is also decoded layer by layer. The difference is that the layer is oriented diagonally rather than horizontally. Consider a DLST codeword matrix **C**, as shown in Fig. 5-2. The entries of **C** below the first NW-SE diagonal are zero. The entries on the first diagonal are thus the undetected symbols of the highest row numbers in their respective columns. To decode the DLST code, the receiver first generates a decision variable for each of the entries on the first diagonal. These decision variables are used by a corresponding 1-D decoder to decode this diagonal. The decision is then fed back to remove the contribution of these symbols to the received signal. The receiver then continues to decode the next diagonal.

Suppose that the rate of the constituent code is fixed, regardless of the number of antennas. The HLST codes and DLST codes obviously offer an overall data rate proportional to n. The decoding complexity of LST codes includes two contributions. The complexity of the spatial processing is on the order of $O(n^2 + nm)$ operations per transmitted n-tuple if linear operations, such as those described in the next section, are employed. The complexity of decoding the n constituent codes can be estimated to be n times the complexity of decoding a typical constituent code. Clearly LST decoding requires much less complexity than ML decoding.

5.4 Error-Probability Analysis For Layered Space-Time Codes

5.4.1 Expressions for the Decision Variables

In the previous section, we have shown that the LST decoding consists of three steps: interference suppression, constituent code decoding, and interference cancellation. There are many schemes that can be used for interference suppression. The choice of interference suppression scheme will affect the decoding performance. In this chapter, we focus on linear zero-forcing (ZF) interference suppression because it leads to a tractable analysis. At high SNR and large n, the performance with linear ZF interference suppression is very close to that with linear MMSE interference suppression [39].

The mathematical formulation of linear zero-forcing (ZF) interference suppression is as follows. We focus on a given instance of time, and we drop the time index for simplicity. Let the QR decomposition of the channel H be $H = UR$, where U is a unitary matrix and R is an upper triangular matrix. We left-multiply the received signal r by $U^\dagger$ to obtain an m-tuple y,

$$y = Ur = Rc + v', \qquad (5\text{-}4)$$

where $v' = U^\dagger v$ is an m-tuple of i.i.d. $\tilde{N}(0, N_0)$ noise components. Because R is upper triangular, for any given row number k, $k = 0, 1, ..., n - 1$,

$$y^k = R_k^k c^k + \{\text{a linear combination of } c^{k+1}, c^{k+2}, ..., c^{n-1}\} + v'^k. \quad (5\text{-}5)$$

The interference term in (5-5) is independent of $c^0, c^1, ..., c^{k-1}$. That is, the interference from these symbols is suppressed. We can remove the interference term in y^k to obtain the decision variable z^k for c^k using the decisions on $c^{k+1}, c^{k+2}, ..., c^{n-1}$. Assuming that these decisions are all correct, z^k is

$$z^k = R_k^k c^k + v'^k, \; k = 0, 1, ..., n-1. \tag{5-6}$$

The power gain $\left|R_k^k\right|^2$ is a quantity that depends on the random channel realization. It has been shown that, if the channel matrix H has i.i.d. circularly symmetric complex Gaussian entries, $\left|R_k^k\right|^2$, $k = 0, 1, ..., n-1$, are independently chi-squared distributed with $2(m-k)$ degrees of freedom [28]. This is an important result that will be used repeatedly in our analysis.

The relationship between c^k and z^k in (5-6) can be interpreted as the input-output relation of a SISO channel with power gain $\left|R_k^k\right|^2$ and AWGN. Because the gains $\left|R_k^k\right|^2$ are independently chi-squared distributed with $2(m-k)$ degrees of freedom, one can interpret (5-6) as the transmission of a symbol c^k over a $(1, m-k)$ receive diversity system to form the decision variable z^k with the use of maximal-ratio combining [40]. This implies an intuitive interpretation that, assuming there are no errors in the feedback of symbol decisions, the kth row of an LST codeword matrix is transmitted over a $(1, m-k)$ system without interference from the other rows of the codeword matrix, and all fades are i.i.d.

5.4.2 Performance of HLST Codes

Consider the kth row of an HLST codeword C. Let $\{c_\tau^k\}$ denote the actual transmitted symbol sequence on this row, and $\{e_\tau^k\}$ denote a distinct possible transmitted symbol sequence. Conditioned on the channel realization $H_\tau = \{H_0, H_1, ...\}$, the probability that the likelihood of transmitting $\{e_\tau^k\}$ is higher than $\{c_\tau^k\}$ is

$$Prob(c^k \rightarrow e^k | H_\tau) = Q\left(\sqrt{\frac{E}{2N_0}\sum_\tau \left|(R_k^k)_\tau\right|^2\left|c_\tau^k - e_\tau^k\right|^2}\right)$$

$$\leq \exp\left\{-\frac{E}{4N_0}\sum_\tau \left|(R_k^k)_\tau\right|^2\left|c_\tau^k - e_\tau^k\right|^2\right\}, \tag{5-7}$$

where the matrix R_τ comes from the QR decomposition of H_τ, i.e.,

$H_\tau = U_\tau R_\tau$. This is the conditional pairwise error probability between $\{c_\tau^k\}$ and $\{e_\tau^k\}$. The average pairwise error probability can thus be upper-bounded by taking the expected value of the right side of (5-7) over the distribution of $\left|(R_k^k)_\tau\right|^2$, which is a chi-squared distribution with $2(m-k)$ degrees of freedom. In a fast fading environment, the $\left|(R_k^k)_\tau\right|^2$ are i.i.d. for distinct τ. The average pairwise error probability, $Prob(c^k \to e^k)$, can be upper bounded by:

$$Prob(c^k \to e^k) \le \prod_{\tau \in \eta(c^k, e^k)} E\left(\exp\left\{-\frac{E}{4N_0}\left|(R_k^k)_\tau\right|^2\left|c_\tau^k - e_\tau^k\right|^2\right\}\right)$$

$$= \prod_{\tau \in \eta(c^k, e^k)} \left(1 + \left|c_\tau^k - e_\tau^k\right|^2 \frac{E}{4N_0}\right)^{-(m-k)}$$

(5-8)

where $\eta(c^k, e^k) = \{\tau \mid c_\tau^k \ne e_\tau^k\}$. In a slow fading environment, $\left|(R_k^k)_\tau\right|^2 = \left|R_k^k\right|^2$ for all τ, and $Prob(c^k \to e^k)$ can be upper bounded by:

$$Prob(c^k \to e^k) \le E\left(1 + \frac{E}{4N_0}\left|R_k^k\right|^2 \sum_\tau \left|c_\tau^k - e_\tau^k\right|^2\right)^{-(m-k)}$$

$$= \left(1 + \frac{E}{4N_0}|c - e|^2\right)^{-(m-k)}$$

(5-9)

5.4.3 Performance of DLST Codes

DLST codes are decoded diagonal by diagonal. Here we consider the probability of a diagonal decision error. Consider the first diagonal of a DLST codeword. On this diagonal, the transmitted symbols are c_τ^τ, $\tau = 0, 1, ..., n-1$. The probability that, under the DLST decoding algorithm, the likelihood of a distinct diagonal $e = \{e_0^0 \, e_1^1 \, ... \, e_{n-1}^{n-1}\}$ is higher than that of the transmitted diagonal $c = \{c_0^0 \, c_1^1 \, ... \, c_{n-1}^{n-1}\}$, conditioned on the channel realization $H_\tau = \{H_0, H_1, ...\}$, is

$$Prob(c \to e | H_\tau) = Q\left(\sqrt{\frac{E}{2N_0} \sum_{\tau=0}^{n-1} \left|(R_\tau^\tau)_\tau\right|^2 \left|c_\tau^\tau - e_\tau^\tau\right|^2}\right)$$

$$\leq \exp\left\{-\frac{E}{4N_0} \sum_{\tau=0}^{n-1} \left|(R_\tau^\tau)_\tau\right|^2 \left|c_\tau^\tau - e_\tau^\tau\right|^2\right\}. \tag{5-10}$$

Equation (5-10) applies in both fast and slow fading environments because the $\left|(R_\tau^\tau)_\tau\right|^2$ are i.i.d. for $\tau = 0, 1, \ldots, n-1$.

The upper bound of the average pairwise error probability is again obtained by taking the expected value of the right-hand side of (5-10). When the SNR is high,

$$Prob(c \to e) \leq \prod_{\tau \in \eta(c,e)} \left(1 + \left|c_\tau^\tau - e_\tau^\tau\right|^2 \frac{E}{4N_0}\right)^{-(m-\tau)}$$

$$\approx \left\{\prod_{\tau \in \eta(c,e)} \left(\left|c_\tau^\tau - e_\tau^\tau\right|^2\right)^{-(m-\tau)}\right\} \left(\frac{E}{4N_0}\right)^{-\sum_{\tau \in \eta(c,e)} m-\tau}, \tag{5-11}$$

where $\eta(c,e) = \{\tau | c_\tau^\tau \neq e_\tau^\tau\}$. When the SNR is low, i.e. $\left|c_\tau^\tau - e_\tau^\tau\right|^2 (E/(4N_0)) \ll 1$ for all τ, because $(1 + mx)^{-1} \approx (1+x)^{-m}$ for small product mx, (5-10) can be approximated by

$$Prob(c \to e) \leq \prod_{\tau \in \eta(c,e)} \left[\left(1 + \varepsilon\frac{E}{4N_0}\right)^{\varepsilon^{-1}}\right]^{-\{|c_\tau^\tau - e_\tau^\tau|^2(m-\tau)\}}$$

$$= \left[\left(1 + \varepsilon\frac{E}{4N_0}\right)^{\varepsilon^{-1}}\right]^{-\left\{\sum_{\tau \in \eta(c,e)} |c_\tau^\tau - e_\tau^\tau|^2(m-\tau)\right\}}, \tag{5-12}$$

where ε is an arbitrarily small positive number. In Appendix II at the end of this chapter, we provide an exact calculation of the average pairwise error probability. A new definition of Q-function is used so that bounding the conditional error probability with the Chernoff bound $Q(x) \leq \exp(-x^2/2)$ is not necessary.

5.4.4 Performance Comparison: DLST vs. HLST

When the performance of HLST and DLST in a slow fading environment are compared, we identify the major shortcoming of the HLST architecture. Let us compare (5-9) and (5-11) in the high SNR regime. For an HLST code, the average pairwise error probability of the bottommost row is inversely proportional to the $(m - n + 1)$th power of SNR. In contrast, in DLST, the average pairwise diagonal error probability between two diagonals c and e is inversely proportional to the $(\sum_{\tau \in \eta(c, e)} m - \tau)$th power of SNR. Therefore, if constituent codes of equivalent throughput and complexity are deployed, the error probability of a DLST code in a slow fading environment can be much lower than that of an HLST code.

The difference between the performance of DLST and HLST can be explained in an intuitive fashion. We have shown that, under the LST architecture, the rows of a codeword matrix can be thought of as being individually transmitted over $(1, m - k)$ diversity reception systems with independent fading. For an HLST code, the output from a constituent code occupies a particular row, thus only uses one of these virtual diversity reception systems. The constituent code transmitted using the bottommost row experiences only $(m - n + 1)$ receive diversity and could become the performance bottleneck. On the other hand, in the DLST architecture, the output from a constituent code is transmitted over the n virtual diversity reception systems in turn. Because the fades associated with these virtual systems are independent, utilizing these systems in turn provides another form of diversity.

Another advantage of DLST over HLST is that in DLST the constituent codes can be just the same code. By contrast, in HLST, because the orders of receive diversity experienced by different constituent codes are different, to efficiently utilize the benefit of receive diversity, lower-rate codes must be used for the lower rows and higher rate codes for the upper rows. In light of these advantages, thereafter in this chapter we will only consider DLST codes.

5.4.5 DLST Codes: Design Criteria and Trade-offs

We propose the following design criteria for DLST codes, making use of equations (5-11) and (5-12).

- We define the truncated multi-dimensional effective length (TMEL) and the truncated multi-dimensional product distance (TMPD) between two distinct diagonals c and e as

$$TMEL \equiv \sum_{\tau \in \eta(c, e)} m - \tau \quad \text{and} \quad TMPD \equiv \prod_{\tau \in \eta(c, e)} \left| c_\tau^\tau - e_\tau^\tau \right|^{2(m-\tau)}. \quad (5\text{-}13)$$

At high SNR, the pairwise error probability between c and e is approximated by $Prob(c \to e) \approx (\text{TMPD})^{-1}(E/4N_0)^{-\text{TMEL}}$. The code design criterion is to maximize the minimum value of $(\text{TMPD})^{-1}(E/4N_0)^{-\text{TMEL}}$ over all pairs of distinct diagonals. If the exact operating SNR is not known but can be assumed to be reasonably high, a good design criterion is to maximize the minimum two-tuple (TMEL, TMPD) in dictionary order.

- At low SNR, the pairwise error probability is approximated by (5-12). We define the exponent $\sum_{t \in \eta(c, e)} \left| c_t - e_t \right|^2 (m - \tau)$ to be the truncated multi-dimensional weighted Euclidean distance (TMED) between c and e. The code design criterion at low SNR is to maximize the minimum TMED between any pair of distinct diagonals.

In the paragraphs above, the word truncated is employed to make explicit one important limitation of DLST codes. To decode a DLST code diagonal by diagonal, the output symbol sequence of a constituent code is decoded block by block. It is desirable to have the number of symbols contained in a diagonal large. To achieve this, one can employ diagonals that are multiple symbol in width. This is illustrated in Fig. 5-3. With a diagonal width of W symbols, a diagonal can contain nW symbols. If $W \neq 1$, the expressions in (5-10) − (5-12) and the design criteria above must be modified accordingly. An important issue is that the use of wider diagonals does not necessary guarantee better performance. One must optimize the constituent code for each value of W.

At high SNR, the pairwise error probability between c and e is approximately inversely proportional to $(E/4N_0)^{-\text{TMEL}}$. The highest achievable TMEL is $\sum_{\tau = 0}^{n-1} m - \tau = (2m + 1 - n)n/2$. Usually this highest achievable TMEL is quite large, and it sometimes is desirable to employ a lower minimum TMEL to allow for higher throughput. The code design trade-off among

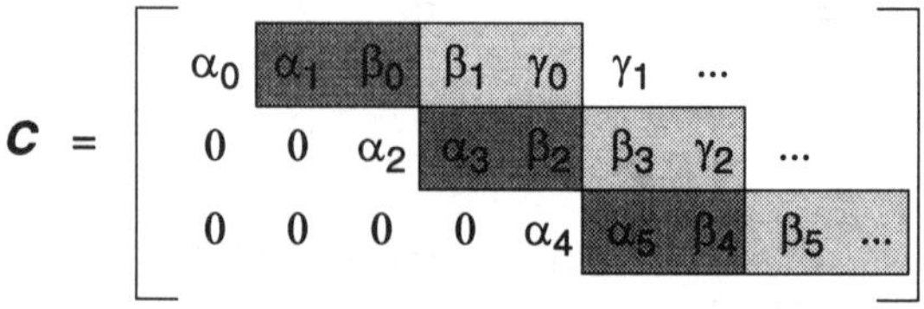

$$C = \begin{bmatrix} \alpha_0 & \alpha_1 & \beta_0 & \beta_1 & \gamma_0 & \gamma_1 & \cdots & & \\ 0 & 0 & \alpha_2 & \alpha_3 & \beta_2 & \beta_3 & \gamma_2 & \cdots & \\ 0 & 0 & 0 & 0 & \alpha_4 & \alpha_5 & \beta_4 & \beta_5 & \cdots \end{bmatrix}$$

Fig. 5-3. DLST with a diagonal width of two symbols.

the constellation size, diagonal width, data rate, and TMEL is as follows. Suppose that the symbol constellation size is 2^b and the diagonal width is W. We claim that, in slow fading environments, it is possible to find a constituent code of data rate R bits/symbol so that the minimum TMEL is at least x:

$$R = \frac{\log_2\{A_{2^{bw}}(n, D)\}}{nW}, \tag{5-14}$$

where $A_{2^{bw}}(n, D)$ is the maximum size of a code length n and minimum Hamming distance D defined over an alphabet of size 2^{bW}, and D is defined by

$$\sum_{\tau = 1}^{D} m - n + \tau \geq x. \tag{5-15}$$

Proof. Denote the symbol constellation by Q. Consider an nW-symbol vector, say $c^{nw} = [c_0 \, c_1 \, \dots \, c_{nW \, 1}]$, whose components (symbols) are defined over Q. Consider a mapping $f: Q^{nW} \to (Q^W)^n$ that maps c^{nw} into a length n vector $(c^W)^n$ whose components are defined over Q^W; $(c^W)^n = [(c_0 \, c_1 \, \dots \, c_{W-1})$ $(c_W \, c_{W+1} \, \dots \, c_{2W-1}) \, \dots \, (c_{(n-1)W} \, c_{(n-1)W+1} \, \dots \, c_{nW-1})]$. The relationship between c^{nw} and $(c^W)^n$ is one-to-one and onto. Let CB denote a set (codebook) of length-n vectors defined over Q^W. According to the input message, the constituent coder chooses a vector $(c^W)^n$ from CB, and then applies f^{-1} on $(c^W)^n$ to obtain c^{nw} and hence the nW symbols to fill a diagonal of width W.

If the minimum Hamming distance of CB is D, clearly the minimum TMEL of the DLST code is at least $(m - n + 1) + (m - n + 2) + \dots + (m - n + D) = \sum_{\tau = 1}^{D} m - n + \tau$. Therefore, it is possible to find a constituent code with data rate $R = \log_2\{A_{2^{bw}}(n, D)\}/nW$ such that the minimum TMEL is at least $\sum_{\tau = 1}^{D} m - n + \tau$. ∎

It is important to know the performance degradation incurred by LST decoding when compared to ML decoding. Consider two DLST codeword matrices C and E. Denote the first diagonals of C and E by c and e, respectively. Further assume that $c \neq e$ and that this is the only difference between C and E. For simplicity, we only consider $W = 1$. The average probability of decoding C as E using ML decoding is given by (5-3), which is equivalent to (5-2) in this special case. The corresponding probability for LST decoding is given by (5-11). Hence:

$$Prob(C \rightarrow E) \leq \prod_{\tau \in \eta(c,e)} \left(1 + \left|c_\tau - e_\tau\right|^2 \frac{E_{ML}}{4N_0}\right)^{-m}$$

for ML decoding and   (5-16)

$$Prob(C \rightarrow E) \leq \left\{ \prod_{\tau \in \eta(c,e)} (\left|c_\tau - e_\tau\right|^2)^{-(m-\tau)} \right\} \left(\frac{E_{DLST}}{4N_0}\right)^{-\sum_{\tau \in \eta(c,e)} m-\tau}$$

for DLST decoding.   (5-17)

It is reasonable to estimate the power penalty by assuming that the (c, e) diagonal pair yields the minimum TMEL. Suppose that the performance requirement is that the average pairwise error probability must not exceed $10^{-\alpha}$. By setting the RHS of both (5-16) and (5-17) to be $10^{-\alpha}$, the power penalty can be estimated to be $\alpha(1 - \text{TMEL}/m|\eta(c, e)|)$ dB.

For example, let us consider a DLST code that achieves the highest achievable TMEL. If $m = n$, $\text{TMEL}/m|\eta(c, e)| \approx 1/2$. To achieve a maximum average pairwise error probability of 10^{-6}, the estimated power penalty is 3 dB. On the other hand, if $m \gg n$, $\text{TMEL}/m|\eta(c, e)| \approx 1/2$. In this case, the power penalty is negligible.

5.5 Operational Aspects of LST Codes

5.5.1 Convolutional Codes as the Constituent Codes

The original DLST architecture does not lend itself to viable implementation using convolutional codes as the constituent codes. Under the DLST architecture, to supply the symbol decisions necessary for interference cancellation, the receiver must determine the values of the symbols of a DLST codeword in a diagonal by diagonal fashion. A diagonal contains a block of nW symbols, and this block is a section of the output symbol sequence of the corresponding constituent (1−D) convolutional code. Diagonal-by-diagonal operation thus means that the output symbol sequence of the constituent code is first partitioned into sections of nW symbols, and each section is decoded without any observations of the subsequent sections. Such a truncation results in poor decoding performance. We can visualize this problem in Fig. 5-2. The symbols α_0, α_1, and α_2 must be determined before observing symbols α_3, α_4, etc. The reliability of the decision on symbol α_2 is particularly in doubt.

To solve this problem, we propose the single-stream structure, which is a modification of the DLST architecture. Under the single-stream structure, only one convolutional constituent code is used. The information bit sequence is fed into the constituent code to generate a coded symbol sequence. The space-time matrix codeword is obtained by filling the NW-SE diagonals of the matrix codeword from left to right with these coded symbols, just as in the original DLST architecture. The single-stream structure is shown in Fig. 5-4.

We will use the following terminology in discussing the decoding procedure. Note that in order to perform decision feedback and interference cancellation, it is only necessary to determine the values of the symbols, not the encoded bits behind those symbols. We refer to a decision on the value of a

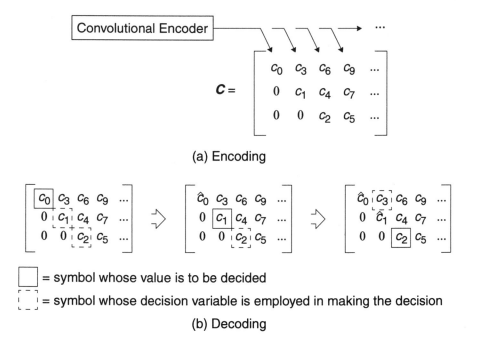

Fig. 5-4. The single stream structure. (a) Encoding. (b) Making symbol-by-symbol tentative decision.

symbol, which is to be used in decision feedback and interference cancellation, as a tentative decision. The decision depth associated with a tentative symbol decision is defined as the number of decision variables corresponding to the symbol to be detected and the subsequent symbols that are incorporated in arriving at the tentative decision. The problem with using convolutional codes as the constituent codes in the original DLST architecture can be rephrased by saying that there are always coded symbols whose tentative decisions are made with excessively low decision depths, i.e., 1, 2, ... symbols.

The slightly modified encoding process in the single-stream structure permits the receiver to perform the task of decision variable generation, tentative decision, and interference cancellation in a symbol-by-symbol manner instead of in a diagonal-by-diagonal manner. It is illustrated in Fig. 5-4. The receiver starts by generating the decision variables for symbols c_0, c_1, c_2, and c_3. The tentative decision on c_0 is first made with a decision depth of 4 (symbols), and then the tentative decision on c_1 is made with a decision depth of 3. With the tentative decision on c_1 available, the decision variable for c_4 can be generated before determining the value of c_2. Now the tentative decision on c_2 can be made with a decision depth of 3, and so on. The benefit is that the decision depth can be at least $W(n-1)$ for every tentative decision. The name "single stream" reflects the use of only one constituent code; by comparing Fig. 5-1 and Fig. 5-4, it is clear that such a symbol-by-symbol decision mechanism is not possible with multiple constituent codes.

The average error probability of tentative decisions can be derived using techniques similar to those employed in deriving (5-10). In the following, we assume a diagonal width $W = 1$; the result can be easily extended to cases where $W > 1$. Consider making a tentative decision on a symbol in the rth row. Suppose that the transmitted symbol is c_0, and the subsequent (sequentially down the diagonal) $n-2$ transmitted symbols are $c_1, ..., c_{n-2}$. Note that c_τ is placed in a slot on the $((\tau + r) \bmod n)$-th row of the space-time codeword matrix. Let another length-$(n-1)$ symbol sequence that can be transmitted in place of $c = \{c_0, c_1, ..., c_{n-2}\}$ be $e = \{e_0, e_1, ..., e_{n-2}\}$. The average probability that, based on the decision variables corresponding to symbols $c_0, c_1, ..., c_{n-2}$, the likelihood of e is higher than that of c, thus resulting in tentatively deciding e_0 instead of c_0, is

$$Prob(c_0 \rightarrow e_0) = E[Prob(c_0 \rightarrow e_0 | H_\tau)]$$

$$\leq \prod_{\tau=0}^{n-2} \left(1 + |c_\tau - e_\tau|^2 \frac{E}{4N_0}\right)^{-(m-row_r(\tau))}, \qquad (5\text{-}18)$$

where $row_r(\tau) = (\tau + r) \bmod n$. Equation (5-18) applies to both fast fading and slow fading environments. At high SNR, equation (5-18) can be approximated as:

$$Prob(c_0 \rightarrow e_0) \leq \left\{ \prod_{\tau \in \eta(c,e)} (|c_\tau - e_\tau|^2)^{-(m-row_r(\tau))} \right\} \times$$

$$\left(\frac{E}{4N_0}\right)^{-\sum_{\tau \in \eta(c,e)} m - row_r(\tau)}, \qquad (5\text{-}19)$$

where $\eta(c, e) = \{\tau | \tau = 0, 1, ..., n-2, c_\tau \neq e_\tau\}$. At low SNR, equation (5-18) can be approximated as:

$$Prob(c_0 \rightarrow e_0) \leq \left[\left(1 + \varepsilon\frac{E}{4N_0}\right)^{-1/\varepsilon}\right]^{\sum_{\tau=0}^{n-2} |c_\tau - e_\tau|^2(m-row_r(\tau))}. \qquad (5\text{-}20)$$

From (5-18) and (5-19), we propose the code design criteria for single-stream DLST codes that minimizes the maximum average pairwise tentative decision error probability. In the following, c and e are two symbol sequences of length $n - 1$ symbols that can be generated from the same state of the encoder of the constituent convolutional code. The design criterion is to design the constituent convolutional code such that the following quantity is maximized:

- $$\min_{r,\, c_0 \neq e_0} \prod_{\tau=0}^{n-2} \left(1 + |c_\tau - e_\tau|^2 \frac{E}{4N_0}\right)^{m-row_r(\tau)} \quad \text{(high SNR regime)},$$

- $$\min_{r,\, c_0 \neq e_0} \sum_{\tau=0}^{n-2} \{m - row_r(\tau)\}|c_\tau - e_\tau|^2 \quad \text{(low SNR regime)}.$$

Error propagation is a common problem in all systems that employ decision feedback. In the LST architecture, errors can propagate both spatially and temporally. Consider a symbol and an incorrect tentative decision on it.

Because the incorrect decision is used in the interference cancellation process, the decision variables of the symbols that are above this symbol on the same column will contain error (errors propagate spatially upward). These incorrect decision variables may further result in more erroneous tentative decisions (errors propagate temporally).

After the tentative decisions are made, a sequence of decision variables is generated. A decoding process is then applied over this decision variable sequence to determine the encoded information bits, which we refer to as the final decisions. The decoding processes for making tentative (symbol) decisions and final (bit) decisions are conceptually separate. The final decisions can be made without being subject to the limitation that the minimum decision depth cannot exceed $W(n - 1)$. One might argue that by using a decision depth longer than $W(n - 1)$, the error probability of the final decisions can be lower than that of the tentative decisions, especially when the minimum span of an error event of the constituent code is longer than $W(n - 1)$ symbols. However, the exact error analysis of both the tentative and final decisions is difficult due to the error propagation phenomenon.

We observe the impact of error propagation through Monte-Carlo simulations. In the range of SNR of interest, the simulation result indicates that the final decision error probability is always a significant fraction of the tentative decision error probability. Thus our design criterion that optimizes the tentative decision error probability is justified.

Example 1. In this example, the number of antennas is $n = m = 8$. The constituent code for the n-D single stream LST code is to be a rate-1/2 feedforward convolutional code with a constraint length no greater than 10. Each pair of output bits from the constituent encoder is mapped to a symbol point on the QPSK constellation using the Gray code mapping. The throughput of the single-stream DLST code is 8 bits per channel use.

For $W = 1$, we perform an exhaustive search in the code space according to our code design criterion in the high SNR regime. The maximum average pairwise error probability of the optimal constituent code is

$$\prod_{\tau = 0}^{n - 2} \left(1 + |c_\tau - e_\tau|^2 \frac{E}{4N_0}\right)^{-(m - row_r(\tau))} \approx 2^{19}\left(\frac{E}{4N_0}\right)^{-11}.$$ This single-stream

DLST code is referred to as code A. The aforementioned convolutional code is also used as the constituent code for an 8-D single stream LST code with a diagonal width of 2 symbols (code B). The maximum average pairwise error probability of code B is approximately $\frac{2^{16}}{3}\left(\frac{E}{4N_0}\right)^{-10}$. We did not search for the optimum constituent code for code B due to time constraints.

We performed Monte-Carlo simulation to obtain the performance of both code A and code B in slow fading environments. The following parameters and terminology are used.

- For code A, the decision depths used for tentative decisions and final decisions are 7 and 14, respectively.

- For code B, the decision depths used for both tentative decisions and final decisions are 14. Because there are many error events whose spans are less than 14 symbols, a decision depth of 14 for final decisions is sufficient.

- An incorrect tentative decision that follows a long sequence of correct tentative decisions is referred to as a *leading error*. As we have shown in the paragraph above, a leading error can trigger numerous subsequent errors. We record the error propagation statistics.

Fig. 5-5(a) shows the probability of occurrence of leading errors vs. SNR. This is also the tentative decision error probability assuming perfect decision feedback. Specifically, a large number of i.i.d. random channel realizations H are generated. For each H, the average probability of leading error $P_{e|H}$ is obtained from simulation. The average leading error probability $\overline{P}_e$ is the average of $P_{e|H}$. We define outage as an event that the channel H cannot support an average error probability $P_{e|H}$ lower than a required level. Thus, the ten-percent outage leading error probability $P_e^{0.1}$ is the highest error probability such that 10% of the randomly generated channels have average error probability $P_{e|H}$ exceeding $P_e^{0.1}$. As expected, code B outperforms code A in average leading error probability when E/N_0 is below 12 dB because of its lower maximum average pairwise error probability. Code B also outperforms code A in the ten-percent outage leading error probability.

Fig. 5-5(b) shows the percentage of channels on which $P_{e|H}$ is lower than 10^{-2} or 10^{-3}. It shows that, for code A and code B, with SNRs of 11 dB and 10 dB, respectively, more than 90% of the random channel realizations support an average leading error probability $P_{e|H}$ lower than 10^{-3}.

From simulation, we identify the channels conditioned on which the average probability of leading error $P_{e|H}$ is close to the ten-percent outage

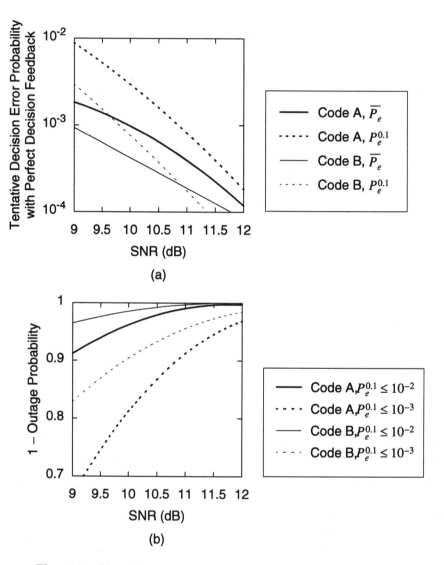

Fig. 5-5. Simulation results of the error performance of single-stream DLST convolutional codes in slow fading environments. (a) Average error probability $\bar{P}_e$ and ten-percent outage probability $P_e^{0.1}$ of tentative decisions assuming perfect decision feedback. (b) Percentages of channels H over which the conditional tentative decision error probability $P_{e|H}$ satisfies the quality requirement assuming perfect decision feedback.

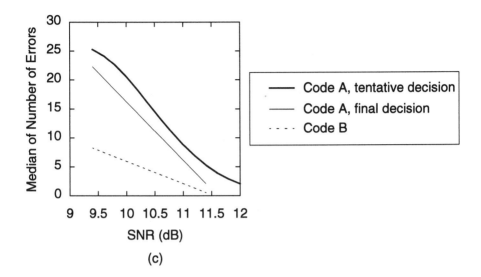

Fig. 5-5. Continued. (c) The median of triggered decision errors over channels conditioned on which the average probability of leading error $P_{e|H}$ is close to the ten-percent outage leading error probability $P_e^{0.1}$.

leading error probability $P_e^{0.1}$. Fig. 5-5(c) shows the median number of errors that are triggered by a leading error on those channels. For code A, the number of triggered final decision errors is lower than the number of triggered tentative decisions errors. This can be attributed to the difference in the decision depths employed. We also observe error propagation events that last a very long period of time, sometimes even hundreds of symbol decisions. ∎

5.5.2 Block Codes as the Constituent Codes

It is natural to use block codes as the constituent codes of a DLST code. To fill a NW-SE diagonal of a DLST codeword matrix, a block of information bits is first encoded using a block code, yielding one (or several) block codeword(s). The block codeword(s) is (are) subsequently mapped into a block of Wn symbols. These Wn symbols are then used to fill a diagonal of the DLST codeword matrix.

In the following we present an analysis of the error probability. We will assume that the diagonal width $W = 1$ and that exactly one block codeword is

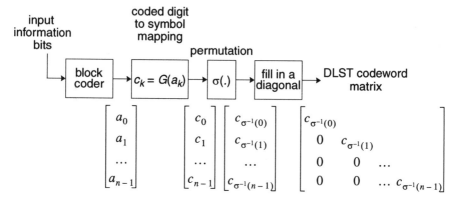

Fig. 5-6. Diagram of DLST encoder using a block code as the constituent code.

contained in a diagonal; the result can be easily generalized to other cases. Let a be a block codeword of n components, $a = (a_0, a_1, ..., a_{n-1})$. Each component of a is individually mapped to a symbol, or a constellation point, under a given mapping; that is, a is mapped into a vector $c = (c_0, c_1, ..., c_{n-1})$ componentwise. These n symbols $c_0, c_1, ..., c_{n-1}$ fill the slots of a diagonal. Specifically, c_k fills the $\sigma(k)$th row slot, where $\sigma(.)$ is a permutation of $0, 1, ..., n-1$. Here, $\sigma(.)$ is called the slot assignment. It is fixed and is known to both the transmitter and the receiver. This encoding process is illustrated in Fig. 5-6. Assuming that there is no preceding decision feedback errors, the average probability that the transmitted vector c is less likely than another vector e is upper-bounded by:

$$Prob(c \rightarrow e) \leq \prod_{k=0}^{n-1} {}^T \left(1 + |c_{\sigma(k)} - e_{\sigma(k)}|^2 \frac{E}{4N_0} \right)^{-(m-k)}, \qquad (5\text{-}21)$$

in both slow and fast fading environments. Because the exponent $m - k$ varies with the row number k, the error probability can depend on the permutation $\sigma(.)$.

Given a slot assignment $\sigma(.)$, the TMEL, TMPD, and TMED between two block codewords c and e are defined in Chapter 5.4(D). The design criteria in Chapter 5.4(D) are applied to calculate the maximum average pairwise error probability given a slot assignment $\sigma(.)$; we then choose the $\sigma(.)$ that yields the lowest maximum average pairwise error probability. In the following, we briefly examine the performance of DLST codes with two classes of popular block codes, the cyclic codes and the linear array codes, as the constituent code.

Example 2. A popular class of block codes is the Reed-Solomon (RS) code. In this example, a (7, 3) RS code is used as a constituent code of a 7-D DLST code. The (7, 3) RS encoder maps three 8-ary input digits into seven 8-ary output digits. Each 8-ary digit selects a point (symbol) on the 8-PSK constellation according to the Gray code mapping, and these seven constellation points are used to fill the slots of a diagonal of the 7-D DLST code.

The RS code is a linear maximal-distance separable (MDS) code. That is, the minimum Hamming distance of an (N, K) RS code is $d^H_{min} = N - K + 1$. We assert that if the block constituent code is a linear MDS code, and if the coded digits of a block codeword are individually mapped to constellation points, all permutations lead to the same minimum TMEL of $d^H_{min}(d^H_{min} + 1)/2 + d^H_{min}(m - n)$.

Proof: Let us denote the inverse function of $\sigma(k)$ by $\sigma^{-1}(k)$. For any slot assignment $\sigma(k)$, consider deleting the $\sigma^{-1}(0)$th, the $\sigma^{-1}(1)$th, ..., and the $\sigma^{-1}(d^H_{min} - 1)$th digits of every codeword of the RS codebook. The resulting $(K - 1, K)$ codebook has zero minimum Hamming distance. Due to linearity, there are at least two all-zero codewords in this new codebook: one is the result of puncturing the original all-zero codeword, and the other is the result of puncturing a weight-d^H_{min} codeword c. When c is transmitted, $c_{\sigma^{-1}(0)}$, $c_{\sigma^{-1}(1)}$, ..., and $c_{\sigma^{-1}(d^H_{min} - 1)}$ fill the slots in the zeroth row to the $(d^H_{min} - 1)$th row, respectively. We have now proved that due to the MDS property, there is always at least one weight d^H_{min} codeword c whose components $c_{\sigma^{-1}(0)}$, $c_{\sigma^{-1}(1)}$, ..., and $c_{\sigma^{-1}(d^H_{min} - 1)}$ are nonzero, and the TMEL between c and the all-zero codeword is $(m - n + 1) + ... + (m - n + d^H_{min})$ and is an upper bound of the minimum TMEL. On the other hand, because the minimum Hamming distance is d^H_{min}, the minimum TMEL is obviously lower-bounded by $(m - n + 1) + ... + (m - n + d^H_{min})$. Therefore, the minimum TMEL is $(m - n + 1) + ... + (m - n + d^H_{min}) = d^H_{min}(d^H_{min} + 1)/2 + d^H_{min}(m - n)$.

Monte-Carlo simulations are performed to obtain the performance of this DLST code in slow fading environments with the number of receiving antennas $m = 7$. Fig. 5-7(a) shows the average error probability assuming perfect decision feedback. Fig. 5-7(b) shows the median number of decision errors measured in number of diagonals that are triggered by a leading error.

■

To achieve the performance indicated by equation (5-21), the receiver needs to perform ML decoding over the decision variables corresponding to the symbols contained in a diagonal. Efficient ML decoding algorithm for RS codes is still unknown. Although the codebook of a (7, 3) RS code is small enough that implementing a brute-force search for ML decoding is still feasible, the need for ML decoding motivates us to study block codes for which there exist efficient ML decoding algorithms.

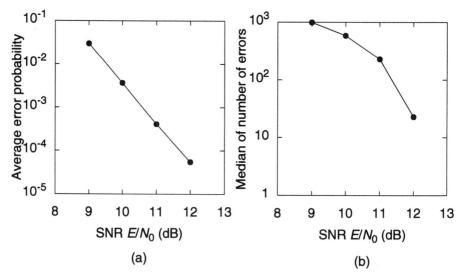

(a)   (b)

Fig. 5-7. Simulation results of the error probability of a 7-D DLST code using the (7, 3) RS as its constituent code in slow fading environments. Here, $n = m = 7$. (a) The average error probability assuming perfect decision feedback. (b) The median number of triggered decision errors over channels conditioned on which the average probability of leading error $P_{e|H}$ is close to the ten-percent outage leading error probability $P_e^{0.1}$.

Example 3. One class of block codes that have compact and regular trellis representations, which permit the use of efficient trellis-based ML decoding, is the linear array codes (LAC) [41].

In this example we consider the following (8, 4) LAC over GF(4). For the encoding, four 4-ary information digits, x_1, x_2, x_3, and x_4, are arranged in a two-by-two matrix:

$$\begin{bmatrix} x_1 & x_2 \\ x_3 & x_4 \end{bmatrix}.$$

For every row and column of this matrix, a parity check digit is generated: $p_1 = x_1 \oplus x_2$, $p_2 = x_3 \oplus x_4$, $p_3 = x_1 \oplus x_3$, and $p_4 = x_2 \oplus x_4$. The four information digits and four parity digits comprise a block codeword c. Gray code mapping is used to map a 4-ary digit to a 4-PSK constellation point. In contrast to the previous example, here the minimum TMEL of the constructed DLST code depends on the slot assignment. With $n = m = 8$, the minimum TMEL can be as high as 13 and as low as 6, depending on the slot assignment used. One of the slot assignments that results in the highest minimum TMEL is shown in Fig. 5-8(a).

Monte-Carlo simulations are performed to obtain the performance of this DLST code in slow fading environments with $m = 8$ and the aforementioned permutation. Fig. 5-8(b) shows the average error probability assuming perfect decision feedback. Fig. 5-8(c) shows the median number of decision errors that are triggered by a leading error. ∎

5.6 Summary

Space-time codes, which embed redundancy in both the temporal and spatial dimensions of the transmitted signal, are channel codes that can be used to utilize the high channel capacity of dual antenna-array systems, particularly for systems in which the transmitter does not have the instantaneous channel state information. In this chapter, we considered space-time codes whose throughput scale linearly in n, assuming that $n \le m$. Accompanying this high throughput, however, is a potentially very high decoding complexity. With

coded symbol	x_1	x_2	x_3	x_4	p_1	p_2	p_3	p_4
fill a slot in row	1	7	6	5	4	3	2	0

(a)

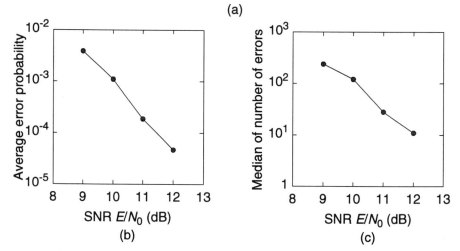

(b)   (c)

Fig. 5-8. Simulation results of the error performance of a 8-D DLST code using the (8, 4) LAC as its constituent code in slow fading environments. Here, $n = m = 8$. (a) One of the permutations that achieves the largest minimum TMEL. (b) The average error probability assuming perfect decision feedback. (c) The median number of triggered decision errors over channels conditioned on which the average probability of leading error $P_{e|H}$ is close to the ten-percent outage leading error probability $P_e^{0.1}$.

ML decoding, the receiver complexity is generally exponential in n, which can be unmanageable even for moderate n.

We proposed LST codes which allow for low-complexity decoding. We first analyzed the performance of LST codes, considering both slow and fast fading environments as well as both high and low SNR situations. From the analysis, we found that in slow fading environments DLST codes outperform HLST codes. We defined the key design parameters – TMEL, TMPD, and TMED – and formulated the design criteria for DLST codes.

We examined the use of convolutional codes and block codes as the constituent codes for DLST codes. With convolutional constituent codes, we introduced the single-stream structure to greatly improve the reliability of tentative decisions. With block constituent codes, we showed that the slot assignment can result in dramatic differences in performance and thus must also be optimized. We formulated the error analysis and design criteria for these modified DLST codes. We also provided example DLST codes and simulated their performances.

Appendix I: ML Decoding Error Analysis

In this appendix, we present an analysis of the error probability achieved by ML decoding, using matrix codeword notation.

Let C and E be two distinct space-time codeword matrices. Suppose that C is the transmitted space-time codeword. Given the channel realizations H_τ, the squared Euclidean distance between the noiseless receptions of C and E is $d^2(C, E|H_\tau) \equiv \sum_\tau |H_\tau(c_\tau - e_\tau)|^2$. The probability that the likelihood given E is transmitted is higher than that given C is transmitted, conditioned on the channel realizations, is

$$Prob(C \rightarrow E|H_\tau) = Q\left(\sqrt{\frac{E}{2N_0} d^2(C, E|H_\tau)}\right)$$

$$\leq \exp\left\{-\frac{E}{4N_0} d^2(C, E|H_\tau)\right\}, \qquad (5\text{-}22)$$

where the Chernoff bound $Q(x) \leq \exp(-x^2/2)$ is applied to form the upper bound. The average pairwise space-time codeword error probability is obtained by averaging equation (5-22) over the distribution of H_τ.

In a fast fading environment, we can define $y_\tau \equiv H_\tau(c_\tau - e_\tau)$ and an $mL \times 1$ vector Y by $Y \equiv (y_1' \ y_2' \ ... \ y_L')'$. Note that $d^2(c, e|H_\tau) = |Y|^2$. Because Y is an mL-dimensional complex Gaussian vector, the eigenvalues of the covariance matrix of Y completely determines the distribution of $|Y|^2$. The covariance matrix of Y is thus simply a block diagonal matrix with the covariance matrices of y_τ on the diagonal:

$$E(YY^\dagger) = diag(E(y_1 y_1^\dagger), E(y_2 y_2^\dagger), ..., E(y_L y_L^\dagger)), \qquad (5\text{-}23)$$

because for fast fading H_τ are mutually independent. Furthermore, because the entries of H_τ are mutually independent, the covariance matrix of y_τ is a diagonal matrix: $E(y_\tau y_\tau^\dagger) = |c_\tau - e_\tau|^2 I_m$. Therefore, $E(YY^\dagger)$ is simply a diagonal matrix, and its diagonal entries are its eigenvalues. Knowing these eigenvalues, the expected value of the right-hand side of (5-22) can be taken to upper-bound the average pairwise error probability $Prob(C \rightarrow E)$ by

$$Prob(C \rightarrow E) \leq \prod_{\tau=1}^{L} \left(1 + |c_\tau - e_\tau|^2 \frac{E}{4N_0}\right)^{-m}. \qquad (5\text{-}24)$$

When analyzing the error performance in slow fading environments, it is easier to consider the transpose of C, $C' = (c^1 \ c^2 \ ... \ c^n)$, where c^k is an m-tuple. Because $H_\tau = H$ for all τ in slow fading, we can drop the time index of H and use h_k to denote the kth column of H. The vector Y can be written in the following form:

$$Y = (c^1 - e^1) \otimes h_1 + (c^2 - e^2) \otimes h_2 + ... + (c^n - e^n) \otimes h_n, \qquad (5\text{-}25)$$

where $\otimes$ denotes the Kronecker product.

The eigenvalues of the covariance matrix of Y is

$$
\begin{aligned}
E(YY^\dagger) &= \sum_{k=1}^{n} \sum_{l=1}^{n} \{(c^k - e^k)(c^l - e^l)^\dagger\} \otimes \{E(h_k h_l^\dagger)\} \\
&= \sum_{l=1}^{n} \{(c^l - e^l)(c^l - e^l)^\dagger\} \otimes I_m \\
&= \{(C - E)'(C - E)^*\} \otimes I_m,
\end{aligned}
\qquad (5\text{-}26)
$$

because $E(h_k h_l^\dagger) = \delta(k-l)I_m$. Let Λ_k denote the nonzero eigenvalues of the matrix $(C - E)(C - E)^\dagger$, $1 \le k \le \text{rank}(C - E)$. Note that the number of nonzero eigenvalues, $\text{rank}(C - E)$, is upper-bounded by $\min(L, n)$. Due to the property of matrix Kronecker product, Λ_k is an eigenvalue of $E(YY^\dagger)$ of multiplicity m. The average pairwise error probability can be upper-bounded by using (5-22),

$$Prob(C \to E) \le \prod_{k=1}^{\text{rank}(C-E)} \left(1 + \Lambda_k \frac{E}{4N_0}\right)^{-m}. \qquad (5\text{-}27)$$

When (5-24) and (5-27) are compared, we find that the upper bound for the average pairwise space-time codeword error probability between C and E is also the upper bound for the average pairwise codeword error probability between two 1-D codewords f and g of length $\text{rank}(C - E)$, $f_k - g_k = \Lambda_k$, transmitted over a $(1, m)$ system in a fast fading environment.

Appendix II: Accurate computation of average pairwise error probability

The Gaussian tail probability function $Q(x)$ is ordinarily defined as

$$Q(x) \equiv \int_x^\infty \frac{1}{\sqrt{2\pi}} \exp\left(-\frac{y^2}{2}\right) dy. \tag{5-28}$$

Craig in his work [44] showed that the Q-function can be defined (but only for $x \geq 0$) by

$$Q(x) \equiv \frac{1}{\pi} \int_0^{\pi/2} \exp\left(-\frac{x^2}{2\sin^2\theta}\right) d\theta. \tag{5-29}$$

In addition to the advantage of having finite integration limits, the form in (5-29) has the argument of the function x in the integrand rather than in the integration limits. This latter property can simplify the exact evaluation of properties of random variables in the form $Q(x)$ wherein x is a random variable of some distribution. Many interesting cases have been solved by Simon and Divsalar [45]. In this chapter, we apply the Chernoff bound, $Q(x) \leq \exp(-x^2/2)$, to approximate the Q-function in order to obtain the error probability. Here, we focus on the exact evaluation of the average pairwise diagonal error probability $Prob(c \rightarrow e)$ for a DLST code.

The conditional pairwise diagonal error probability is specified in (5-10):

$$Prob(c \rightarrow e|H_\tau) = Q\left(\sqrt{\frac{E}{2N_0}\sum_{\tau=0}^{n-1}\left|(R_\tau^\tau)_\tau\right|^2\left|c_\tau^\tau - e_\tau^\tau\right|^2}\right), \tag{5-30}$$

where $\left|(R_\tau^\tau)_\tau\right|^2$, $\tau = 0, 1, \ldots, n-1$, are independently chi-squared distributed with $2(m-\tau)$ degrees of freedom. The average pairwise error probability is thus

$$Prob(c \rightarrow e) = \int_0^\infty\int_0^\infty \cdots \int_0^\infty Q\left(\sqrt{\frac{E}{2N_0}\sum_{\tau=0}^{n-1}\left|(R_\tau^\tau)_\tau\right|^2\left|c_\tau^\tau - e_\tau^\tau\right|^2}\right)$$
$$d\left|(R_0^0)_0\right|^2 d\left|(R_1^1)_1\right|^2 \ldots d\left|(R_{n-1}^{n-1})_{n-1}\right|^2. \tag{5-31}$$

Using (5-29), (5-31) can be reduced to a single integral with finite limits as follows:

$$
Prob(c \to e) = \int_0^\infty \int_0^\infty \cdots \int_0^\infty \frac{1}{\pi} \int_0^{\pi/2} \exp\left(-\frac{\frac{E}{2N_0}\sum_{\tau=0}^{n-1}\left|(R_\tau^\tau)_\tau\right|^2\left|c_\tau^\tau - e_\tau^\tau\right|^2}{2\sin^2\theta}\right) d\theta
$$

$$
d\left|(R_0^0)_0\right|^2 d\left|(R_1^1)_1\right|^2 \cdots d\left|(R_{n-1}^{n-1})_{n-1}\right|^2
$$

$$
= \frac{1}{\pi}\int_0^{\pi/2}\left[\prod_{\tau=0}^{n-1}\int_0^\infty \exp\left(-\frac{\frac{E}{4N_0}\left|(R_\tau^\tau)_\tau\right|^2\left|c_\tau^\tau - e_\tau^\tau\right|^2}{\sin^2\theta}\right)d\left|(R_0^0)_0\right|^2\right]d\theta \quad (5\text{-}32)
$$

$$
= \frac{1}{\pi}\int_0^{\pi/2}\left[\prod_{\tau=0}^{n-1}\left(1 + \frac{\frac{E}{4N_0}\left|c_\tau^\tau - e_\tau^\tau\right|^2}{\sin^2\theta}\right)^{-(m-\tau)}\right]d\theta.
$$

When the SNR is high, i.e. $\dfrac{E}{4N_0}\left|c_\tau^\tau - e_\tau^\tau\right|^2 \gg 1 \geq \sin^2\theta$, (5-32) can be approximated by

$$
Prob(c \to e) \approx \frac{1}{\pi}\int_0^{\pi/2}\left[\prod_{\substack{\tau=0,\\ \tau \in \eta(c,e)}}^{n-1}\left(\frac{E}{4N_0}\left|c_\tau^\tau - e_\tau^\tau\right|^2\right)^{-(m-\tau)}(\sin\theta)^{2(m-\tau)}\right]d\theta
$$

$$
= \left\{\frac{1}{\pi}\int_0^{\pi/2}[\sin^2\theta]^{TMEL}d\theta\right\}(TMPD)^{-1}\left(\frac{E}{4N_0}\right)^{-TMEL}.
$$

(5-33)

Note that in (5-33), the quantity $\left\{\dfrac{1}{\pi}\int_0^{\pi/2}[\sin^2\theta]^{TMEL}d\theta\right\}$ is monotonically decreasing with TMEL. When the SNR is low, (5-32) can be better approximated instead by

$$Prob(c \rightarrow e) \approx \frac{1}{\pi}\int_0^{\pi/2}\left(1 + \frac{\dfrac{E}{4N_0}TMED}{\sin^2\theta}\right)^{-1} d\theta. \tag{5-34}$$

The same design criteria for DLST codes can be derived from (5-33) and (5-34).

6

Transmit Diversity

6.1 Introduction

In applications such as cellular mobile radio and fixed wireless local loop, antenna arrays are deployed at the base station to combat fading, reject interference, and achieve higher antenna directionality. As the number of antenna elements increases, the quality of the reverse link – the transmission from the user equipment to the base station – can become superior to that of the forward link, even though traditionally the forward link signal strength is much higher than that of the reverse link. Improving the capacity of the forward link has indeed become the priority in many applications.

Unfortunately, there are often constraints, such as size, cost, or battery life, which limit the number of antenna elements that can be deployed at the user equipment. To improve the quality of the forward link, forward transmit diversity must be effectively utilized. However, whereas techniques to utilize receive diversity are well documented, transmit diversity was not aggressively explored until recently.

The reason that transmit diversity is less straightforward to employ than receive diversity is that when the number of transmit antennas is greater than the number of receiving antennas, the spatial dimension of the transmit signal is greater than that of the received signal. Thus even when the channel matrix is of full rank, linear operations at the receiver can no longer separate signals transmitted by different antennas. In the past, to avoid making the signal detection task too difficult, it was common to reduce the spatial dimension of the transmit signal through decimation. For example, in switched transmit diversity, only one of the transmit antennas is permitted to transmit at any given time. Other transmit diversity techniques include intentional time offset, frequency offset, phase sweeping, frequency hopping and modulation diversity. Decimation of the spatial degree of the transmitted signal usually results in forfeiting some of the potential of transmit diversity.

In the previous chapters, we focused on systems that have more receive antennas than transmit antennas. In this chapter, we will address the capacity and efficient signal processing algorithms for the case when the number of transmit antennas exceeds that of receiving antennas. In fact, many of the dual antenna array concepts introduced in the previous chapters apply even when $m < n$. In this chapter, we will only concentrate on the aspects specific to the cases where $m < n$.

In this chapter, we first quickly review the channel capacity and the optimal signal processing architecture with multiple transmit diversity from an information-theoretic point-of-view. One particularly interesting result is that when $m = 1$, with uniform power allocation, the capacity distribution of an $(n, 1)$ channel at SNR $n\rho$ is identical to that of a $(1, n)$ channel at SNR ρ. However, effective decoding algorithms to process a transmitted signal with uniform power allocation are still unknown. We will describe transmission techniques that directly lead to an input-output formulation that is similar to that of a $(1, n)$ receive diversity system with maximal ratio combining. These techniques are very attractive because they greatly simplify the receiver task, inasmuch as they cannot always achieve channel capacity. An additional advantage is that they do not require a complete redesign of existing systems and thus they are well suited for upgrading their quality.

The remainder of this chapter is organized as follows. In chapter 6.2, we review the channel capacity of dual antenna-array systems. In particular, we will focus on $m = 1$. In chapter 6.3, we examine a simple method applicable to any $(n = 2, m)$ channel that indeed achieves the channel capacity with uniform power allocation. In chapter 6.4, we introduce the generalization of this technique to $n > 2$. We give the summary in chapter 6.5.

6.2 Channel capacity when the number of transmit antennas exceeds the number of receive antennas

As we have mentioned in chapter 3.3, the channel capacity of an (n, m) channel given the channel realization H subject to an average normalized transmitter power constraint is

$$C = \max_{tr(\Sigma_s) \le \rho} \log_2[\det (I + H\Sigma_s H^\dagger)] \text{ (bits per channel use).} \qquad (6\text{-}1)$$

Denote the singular value decomposition of the channel matrix H be $H = U_H D_H V_H^\dagger$. When the transmitter has CSI, the input that achieves the channel capacity is a zero-mean complex Gaussian vector with autocovariance matrix $V_H D V_H^\dagger$. The channel capacity is

$$C_{\text{opt}} = \sum_{k=1}^{m} \log_2(1 + D_k^k \varepsilon_k^2), \qquad (6\text{-}2)$$

where ε_k^2 is a nonzero eigenvalue of $HH^\dagger$ and the nonnegative diagonal matrix D is the one that maximizes the right-hand side of (6-2) subject to the constraint that $tr(D) \le \rho$. The upper summation limit is m because $m < n$.

When uniform power allocation is employed, the maximum mutual information is

$$C_{\text{uni}} = \sum_{k=1}^{m} \log_2\left(1 + \frac{\rho}{n}\varepsilon_k^2\right). \qquad (6\text{-}3)$$

The case when $m = 1$ is of particular interest to applications involving compact user terminal equipment. When $m = 1$, the channel capacities given CSI at the transmitter and the channel capacity with uniform power allocation are $C_{\text{opt}} = \log_2(1 + \rho HH^\dagger)$ and $C_{\text{uni}} = \log_2(1 + (\rho/n)HH^\dagger)$, respectively. It is interesting to see that with optimal power allocation the required transmit power to achieve a certain channel capacity is $10\log n$ dB lower than that with uniform power allocation.

One of the most straightforward ways of utilizing transmit diversity is to have the transmitting antennas sending their signals in turn. The average channel capacity under this technique is

$$C_{\text{switch}} = \frac{1}{n}\sum_{k=1}^{n} \log_2(1 + \rho|H_k|^2), \qquad (6\text{-}4)$$

where H_k is the channel gain associated with the kth transmitting antenna element. Due to the concave property of the logarithmic function, C_{uni} is always greater than or equal to C_{switch}:

$$
\begin{aligned}
C_{\text{switch}} &= \frac{1}{n}\sum_{k=1}^{n} \log_2(1 + \rho|H_k|^2) \\
&\leq \log_2\left(\frac{\sum_{k=1}^{n} 1 + \rho|H_k|^2}{n}\right) = C_{\text{uni}}.
\end{aligned}
\qquad (6\text{-}5)
$$

A additional disadvantage of switched transmit diversity is that antennas need to be turned on and off. This complicates the design of output amplifiers due to the high peak-to-average requirement.

Another popular transmit diversity technique is select transmit diversity, wherein the transmitter activates only the antenna element that has the highest gain. The capacity with this scheme is

$$C_{\text{select}} = \log_2(1 + \rho \cdot \max_{k} (|H_k|^2)). \qquad (6\text{-}6)$$

Clearly, C_{select} is greater than C_{uni} but lower than C_{opt}.

Table 6-1 compares the four transmit diversity techniques described above. Obviously, if complete CSI is available, the transmitter can use the optimal scheme to achieve C_{opt}. If the phase information of the channel gain is not available but the amplitude information is, select diversity is desirable. Otherwise, the transmitter should seek to achieve C_{uni}.

6.3 Transmit diversity equals two

In this section, we will describe a transmission diversity technique first proposed by Alamouti [46]. This technique is can effectively achieve C_{uni} and is applicable to any $(n = 2, m)$ channel with no CSI available at the transmitter, as long as the channel does not vary at a rate comparable to the baud

CSI available to the transmitter	Transmission technique	Corresponding channel capacity
amplitude and phase of the channel gain	optimal power allocation	$\log_2(1 + \rho \sum_{k=1}^{n} |H_k|^2)$
amplitude only	select the best transmit antenna	$\log_2\left(1 + \frac{\rho}{n} \sum_{k=1}^{n} |H_k|^2\right)$
unavailable	uniform Power allocation	$\log_2(1 + \rho \cdot \max_k (|H_k|^2))$

Table 6-1: The appropriate transmit diversity technique and the corresponding channel capacity for systems with different types of CSI available at the transmitter.

rate. We will focus on the case where $m = 1$. Later we will show that generalization to $m > 1$ is straightforward. The major benefit of this algorithm is that it achieves C_{uni} with a very simple encoding/decoding process. Therefore, it is well suited for the next generation cellular radio standard.

The encoding process is as follows. Immediately before time τ, $\tau = 0, 2, 4,$ ..., two symbols s_τ^0 and s_τ^1 arrive at the transmitter. The transmitter sends s_τ^0 to transmit antenna 0 and s_τ^1 to transmit antenna 1, respectively. At the next time instance, i.e. $\tau = 1, 3, 5,$ ..., no data arrives at the transmitter. The transmitter sends $-s_\tau^{1*}$ and s_τ^{1*} to transmit antenna 0 and transmit antenna 1, respectively.

The input-output relationship of the channel at time τ and $\tau + 1$ is

$$\begin{bmatrix} r_\tau \\ r_{\tau+1} \end{bmatrix} = \begin{bmatrix} h_0 & h_1 & 0 & 0 \\ 0 & 0 & h_0 & h_1 \end{bmatrix} \begin{bmatrix} s_\tau^0 \\ s_\tau^1 \\ -s_\tau^{1*} \\ s_\tau^{0*} \end{bmatrix} + \begin{bmatrix} v_\tau \\ v_{\tau+1} \end{bmatrix}, \tau = 2, 4, 6, \dots . \quad (6\text{-}7)$$

Equation (6-7) can be rearranged as

$$\begin{bmatrix} r_\tau \\ r_{\tau+1}{}^* \end{bmatrix} = H \begin{bmatrix} s_\tau^0 \\ s_\tau^1 \end{bmatrix} + \begin{bmatrix} v_\tau \\ v_{\tau+1}{}^* \end{bmatrix}, \qquad (6\text{-}8)$$

where $H = \begin{bmatrix} h_0 & h_1 \\ -h_1{}^* & h_0{}^* \end{bmatrix}$.

Note that the columns of the matrix H are orthogonal to each other. To detect the transmitted symbol, the receiver simply left-multiplies $\begin{bmatrix} r_\tau & r_{\tau+1}{}^* \end{bmatrix}'$ by $H^\dagger$:

$$H^\dagger \begin{bmatrix} r_\tau \\ r_{\tau+1}{}^* \end{bmatrix} = \begin{bmatrix} y_\tau \\ y_{\tau+1} \end{bmatrix} = \begin{bmatrix} |h_0|^2 + |h_1|^2 & 0 \\ 0 & |h_0|^2 + |h_1|^2 \end{bmatrix} \begin{bmatrix} s_\tau^0 \\ s_\tau^1 \end{bmatrix} + \begin{bmatrix} \eta_\tau \\ \eta_{\tau+1} \end{bmatrix}. \quad (6\text{-}9)$$

This simple operation is akin to maximal-ratio combining. In (6-9), η_τ and $\eta_{\tau+1}$ are i.i.d. $\tilde{N}(0, |h_0|^2 + |h_1|^2)$. After normalization, equation (6-9) describes a SISO channel with channel power gain $|h_0|^2 + |h_1|^2$. If h_0 and h_1 are independent $\tilde{N}(0, 1)$, $|h_0|^2 + |h_1|^2$ is chi-squared distributed with four degrees of freedom.

Under this transmit diversity scheme, given the channel gains h_0 and h_1, the channel capacity is

$$C = \log\!\left(1 + (|h_0|^2 + |h_1|^2)\frac{\rho}{n}\right). \qquad (6\text{-}10)$$

This is exactly equal to C_{uni}, given by (6-3).

6.4 Transmit diversity greater than two

To extend the technique in chapter 6.3 to an $(n > 2, m = 1)$ system, one solution is to apply orthogonal designs [47]. The theory of orthogonal design is beyond the scope of this book. In the following, we will introduce this concept using a simple example. The work of Tarokh *et. al* [47] is recommended to interested readers. First, we introduce the concept of orthogonal design.

Definition. A generalized orthogonal design G of size n and rate $R = k/p$ is a $p \times n$ matrix with entries $0, \pm x_1, \pm x_2, ..., \pm x_k$ such that $G^\dagger G = D$ where D is a diagonal matrix with diagonal D_{ii}, $i = 1, 2, ..., n$, in the form $(l_1^i |x_1|^2 + l_2^i |x_2|^2 + ... + l_k^i |x_k|^2)$. The coefficients are positive integers.

It is shown that, without loss of generality, one can consider only $p \times n$ generalized orthogonal designs G in variables $x_1, x_2, ..., x_k$ that satisfy $G^\dagger G = (|x_1|^2 + |x_2|^2 + ... + |x_k|^2)I$.

An example of generalized complex orthogonal design with rate $R = 1/2$ is

$$
G_c^3 = \begin{bmatrix}
x_1 & x_2 & x_3 \\
-x_2 & x_1 & -x_4 \\
-x_3 & x_4 & x_1 \\
-x_4 & -x_3 & x_2 \\
x_1^* & x_2^* & x_3^* \\
-x_2^* & x_1^* & -x_4^* \\
-x_3^* & x_4^* & x_1^* \\
-x_4^* & -x_3^* & x_2^*
\end{bmatrix}. \tag{6-11}
$$

Transmission Scheme. Consider a system with n transmitting antennas and 1 receiving antenna. The encoder first decides on an appropriate rate $R = k/p$ such that a complex generalized orthogonal design of size n, i.e. a $p \times n$ matrix G using k independent symbol, exists. Immediately before $\tau = 0$, k symbols $x_1, x_2, ..., x_k$ arrive at the encoder. The transmitter forms the complex generalized orthogonal design G using these k input symbols. During time $\tau = 1, 2, ..., p - 1$, transmit antenna j transmits the $(\tau+1)$–th row, jth column entry of G. The receiver, upon reception of signals $r_1, r_2, ..., r_p$, performs a linear, matched filtering-like processing to obtain interference-free observations of $x_1, x_2, ..., x_k$.

We will illustrate this scheme using an example. Suppose that $n = 3$ and that the complex generalized orthogonal design in (6-11) with $p = 8$ and $k = 4$ is employed. Denote the channel (complex) gain from transmitting antenna 1, 2, 3 to the receiving antenna by α, β, and γ. The received signals, $r_0, \quad r_1, \quad \ldots, \quad r_7, \quad$ are $\quad r_0 = \begin{bmatrix} \alpha & \beta & \gamma \end{bmatrix} \begin{bmatrix} x_1 & x_2 & x_3 \end{bmatrix}' + v_0$, $r_1 = \begin{bmatrix} \alpha & \beta & \gamma \end{bmatrix} \begin{bmatrix} -x_2 & x_1 & -x_4 \end{bmatrix}' + v_0, \quad \ldots, \quad r_7 = \begin{bmatrix} \alpha & \beta & \gamma \end{bmatrix} \begin{bmatrix} -x_4^* & -x_3^* & x_2^* \end{bmatrix}' + v_7$, respectively.

Define the following vectors:

$$r = \begin{bmatrix} r_0 & r_1 & r_2 & r_3 & r_4^* & r_5^* & r_6^* & r_7^* \end{bmatrix}',$$

$$x = \begin{bmatrix} x_0 & x_1 & \ldots & x_{k-1} \end{bmatrix}', \text{ and}$$

$$v = \begin{bmatrix} v_0 & v_1 & v_2 & v_3 & v_4^* & v_5^* & v_6^* & v_7^* \end{bmatrix}'.$$

It can be easily verified that

$$r = Hx + v, \tag{6-12}$$

where

$$H = \begin{bmatrix} \alpha & \beta & \gamma & 0 \\ \beta & -\alpha & 0 & -\gamma \\ \gamma & 0 & -\alpha & \beta \\ 0 & \gamma & -\beta & -\alpha \\ \alpha^* & \beta^* & \gamma^* & 0 \\ \beta^* & -\alpha^* & 0 & -\gamma^* \\ \gamma^* & 0 & -\alpha^* & \beta^* \\ 0 & \gamma^* & -\beta^* & -\alpha^* \end{bmatrix}. \tag{6-13}$$

Note that the columns of H are orthogonal to each other. More specifically, $HH^\dagger = 2(|\alpha|^2 + |\beta|^2 + |\gamma|^2)I_4$. Because of this property, the receiver can pre-multiply r by $H^\dagger$ to obtain the individual interference-free observations of the transmitted symbols. That is,

$$H^{\dagger} r = \begin{bmatrix} y_0 \\ y_1 \\ y_2 \\ y_3 \end{bmatrix} = 2(|\alpha|^2 + |\beta|^2 + |\gamma|^2) \begin{bmatrix} x_0 \\ x_1 \\ x_2 \\ x_3 \end{bmatrix} + \begin{bmatrix} \eta_0 \\ \eta_1 \\ \eta_2 \\ \eta_3 \end{bmatrix}, \qquad (6\text{-}14)$$

where η_k, $k = 0, 1, 2, 3$, are i.i.d. $\tilde{N}(0, 2(|\alpha|^2 + |\beta|^2 + |\gamma|^2))$. Writing (6-14) componentwise, we have

$$y_l = 2(|\alpha|^2 + |\beta|^2 + |\gamma|^2)x_l + \eta_l. \qquad (6\text{-}15)$$

The signal-to-noise ratio of each symbol observation y_l is $2(|\alpha|^2 + |\beta|^2 + |\gamma|^2)\rho/n$.

Equation (6-14) is identical to the describes a $(1, n)$ system in which a rate-$(1/2)$ repetition code and maximal ratio combining is employed. In general, when a generalized complex orthogonal design of rate R is employed, the channel capacity with this technique is

$$R\log_2\left(1 + \frac{\rho}{Rn}\sum_{k=1}^{n}|H_k|^2\right). \qquad (6\text{-}16)$$

The existence of a generalized complex orthogonal design is key to this technique. It has been shown that complex generalized orthogonal design exists whenever $R \le 0.5$.

6.5 Summary

In this chapter we presented theories and algorithms for channels in which the number of transmitting antennas is higher than the number of receiving antennas.

When the transmitter has the full CSI, it can employ optimal power allocation. In this scenario, the transmitted vector is the linear combination of m orthornormal vectors.

When the transmitter knows only the amplitude gains from the transmitting antenna elements to the single receiving antenna element, a good strategy is to transmit only from the antenna with the highest amplitude gain.

If the transmitter is completely unaware of the channel conditions, uniform power allocation achieves a capacity that is identical to the capacity of a $(1, n)$ channel with a power penalty of $10\log_{10}n$ dB. We described an effective technique that achieves this capacity when $n = 2$. It is possible to extend this technique to channels with $n > 2$ using generalized complex orthogonal design. With this technique, the $(n, 1)$ channel is transformed into an equivalent $(1, n)$ channel with a rate R repetition code and a power penalty of $10\log_{10}n$ dB.

7

Open Issues

7.1 Introduction

In this book we have discussed a few facets of dual antenna-array systems. We have examined the distribution of channel capacity for a typical outdoor base station-subscriber unit link, the performance and applicability of various power allocation strategies, and space-time codes and their design criteria.

Nonetheless, there are still many open issues that must be resolved before there is a complete dual antenna-array solution. In this chapter, we identify a few key areas that warrant further research and development. What follows is by no means a complete list.

7.2 Further Understanding of Channel Statistics

In this book, we have largely focused on quasi-static channels using a model approach. "One-ring" and "two-ring" models are proposed to investigate the spatial fading correlation, assuming that the channel stays constant over a burst transmission period and that the bandwidth is narrow. The "one-ring" model is appropriate in situations where one end of the wireless link is elevated and unobstructed by local scatterers and the other end is surrounded by local scatterers, while the "two-ring" model is appropriate in describing peer-to-peer systems in which local scatterers are present at both ends of the link. We have assumed Rayleigh fading throughput; generalizing our methodology to both Rician and Nakagami fading is not difficult.

We have chosen the one- and two-ring models because they represent two prototypical radio environments. When a more precise description of the channel statistics in a particular environment is desired, one must employ a channel model that is tailored for the specific environment [19]. Such a model should, of course, be validated through experimental measurements.

In this book, we did not address the fading correlation for wideband, time-varying channels. Specifically, denote the channel gain between transmitting antenna TA_k and receiving antenna RA_l at frequency f_i and time τ_j by $h(TA_k, RA_l, f_i, \tau_j)$. A full description of channel fading correlation is to specify $E[h(TA_{k_1}, RA_{l_1}, f_{i_1}, \tau_{j_1}) h(TA_{k_2}, RA_{l_2}, f_{i_2}, \tau_{j_2})^\dagger]$. This information will be necessary in designing and evaluating the performance of space-time channel codes in many practical applications.

7.3 Acquisition and Tracking of Channel State Information

We have shown that in order to realize the potential of dual antenna-array systems, the receiver must be able to measure and track the channel. Computationally efficient schemes for estimating and tracking a channel matrix that has nm entries are desirable.

Transmitted reference techniques usually provide the simplest method for CSI estimation. Common transmitted reference techniques are tone-calibration techniques and pilot symbol-assisted modulation. Both can be modified to accommodate dual antenna arrays [36].

It can sometimes be desirable to directly track certain attributes of the channel other than the channel matrix H. For example, as mentioned in chapter 3, the optimal transmit basis can be used at the transmitter to transform the MIMO channel into parallel SISO subchannels. To apply the optimal transmit basis, the singular value decomposition of the channel matrix, i.e., $H = U_H D_H V_H^\dagger$, must be known. As the channel changes, it is computationally inefficient to track H and continually recompute the singular value decomposition, and a better solution is to track the matrices U_H, D_H, and $V_H^\dagger$.

7.4 Signal Processing Techniques

Chapter 4 discussed the performance of dual antenna-array systems with optimum, uniform, and stochastic water-filling power-allocation strategies in both independent and correlated fading environments. Optimum power allocation achieves the highest capacity and the lowest receiver complexity, but it allocates unequal power, and hence assigns unequal throughput, to the subchannels. When using one of the other two power-allocation strategies, it is often desirable to one-dimensionalize the MIMO channel into a set of parallel SISO subchannels. This reduces the receiver complexity, and typically incurs only a small capacity penalty. The communication rates over these parallel SISO subchannels are unequal.

When there is a significant degree of inequality between subchannel throughputs, for each subchannel it is necessary to employ a subchannel-specific modulation format and channel coding scheme. In particular, to apply optimum power allocation over time-varying channels, the modulation formats and channel coding schemes on the subchannels must be updated to reflect the changes in channel conditions.

7.5 Network Issues

Consider a "benchmark" cellular system in which both the base stations and the subscriber units have only one antenna element. By installing n-element smart antennas at the base station sites while continuing to use only one antenna element at each subscriber unit, the Erlang capacity, or user density per cell, of this "smart antenna-enhanced" system can be increased by approximately a factor of n, as a result of the interference suppression and beamforming capabilities of the antenna array.

In this book, the focus is on a single user-to-single user link with antenna arrays at both ends. We have demonstrated that the channel capacity of an (n, n) link is approximately n times higher than that of a $(1, n)$ link assuming independent fading. If all the base stations and subscriber units are all upgraded to use n-element antenna arrays, however, the supported user density does not exhibit an n-fold increase over the smart antenna-enhanced system described in the previous paragraph.

To further understand this apparent paradox, consider the following time-division multiple access scheme. There are n subscribers in a cell, and the base station spends an equal amount of time serving each subscriber. Note that no more than two subscribers can be served at the same time because the available spatial degrees of freedom of an entire n-element antenna array must be dedicated to one base station-to-subscriber unit link. It is easy to see that in this scenario each subscriber receives a throughput slightly better than that in the benchmark cellular system. So is there any significant advantage of installing n-element antenna arrays at the subscriber units, other than the antenna gain?

The answer lies in the fact that, with dual antenna arrays, the peak throughput of any base station-subscriber unit link is approximately n times higher than that in the "benchmark" system. This is a great advantage if the traffic pattern over the network is bursty, which is often the case in packet data networks. By activating a BS-SU link only when it is required, the response time (delay) of the system can be reduced significantly. To take advantage of this flexible throughput-allocation property, the air interface must be designed with this in mind. In contrast, in a smart antenna-enhanced system, even when there is only one subscriber present in a cell, the peak link throughput cannot be made n times higher than that of a $(1, 1)$ link.

7.6 Distributed BS Antenna Scheme

Throughout this book, we have implicitly assumed that the antenna elements belonging to an "antenna array" are located in relative proximity. In chapter 3, the "one-ring" model is used to study the physical separation required between the BS antenna elements to achieve low fading correlation. As a rule of thumb, the larger the separation, the lower the fading correlation. However, if most of the received energy comes through a dominant line-of-sight path between the two antenna arrays, the channel capacity is dominated by this line-of-sight path and does not exhibit a linear growth with respect to the number of antenna elements at a reasonable SNR.

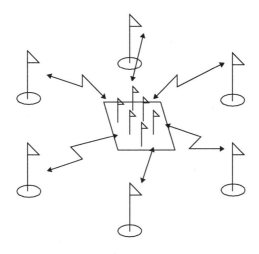

Fig. 7-1. An user with a six-element antenna arrays communicates with six base stations that are far apart from each other.

What happens if the antenna elements are not placed in close proximity to each other? Fig. 7-1 shows an interesting scenario in which a user communicates with six antennas that are placed far apart from each other. In fact, one may find similarity between the scenario shown in Fig. 7-1 with the so-called "soft hand-off" scheme.

It has been shown [23] that in a scenario like Fig. 7-1, the correlation between any two channel fades associated with different distributed BS antenna elements is likely to be low, regardless of whether dominant line-of-sight paths exist between the user antenna array and the distributed BS antennas. Therefore, the probability is high that the channel supports $\min(n,m)$ active spatial modes, with or without fading!

The challenge of operating such a system is obvious. Because the received signal at different distributed sites must be jointly processed, the six base stations must be precisely coordinated. This would require a high-bandwidth communication channel between the controlling entity and the six base stations.

7.7 Space-Time Codes

In chapter 5, we presented a family of codes constructed based on the layered space-time architecture. This particular family of space-time codes can be decoded with a complexity that is only quadratic with the number of antenna elements.

Other high-throughput, low-complexity space-time codes are expected to appear in the near future. One approach that can improve the performance of DLST codes is to utilize an iterative decoding procedure. Iterative decoding for channel codes has received tremendous attention in the communications community recently. Iterative decoding often achieves performance close to the optimal decoding technique, while providing the advantage of greatly reduced complexity. Therefore, it is very attractive for decoding applications of otherwise intractable complexity, such as turbo codes [42].

Some preliminary results [49] indicate that iterative decoding of DLST codes indeed provides a performance improvement over the original hard decision-feedback, diagonal-by-diagonal decoding architecture.

Index

H

Hamming distance 93
HLST 69
Householder transformation 26

I

iterative decoding 118

K

Kronecker product 24, 51

L

layered space-time, LST
  architecture 47, 69
  code 4
  decoding 75
  encoding 73
leading error 89
likelihood 12
linear array code 93

M

maximal distance separable MDS 93
maximal ratio combining 23, 104
MIMO 3, 25
minimum mean squared error, MMSE 60, 77
ML detection 4, 72
multipath 1
multistage decoding 70
mutual information 13

O

one dimensionalization 57
one-ring 28
ordered eigenvalue 33
orthogonal design 108

outage probability 10

P

pairwise error probability 71
power allocation
  optimum 46, 48
  stochastic water-filling 46
  strategy 4, 46
  uniform 46, 50

Q

Q function 78
QR decomposition 26, 51
quasi-static 7

R

rank 99
ray-tracing 18
Reed-Solomon code 93

S

scatterer 2
  actual 20
  effective 20
  model 18
single-stream structure 85
singular value decomposition 25, 48
SISO 4, 25
slot assignment 92
space-time
  channel code 69
  signal processing 3
spatial diversity 1
special complex Gaussian 21
spectral efficiency 1, 4,
stochastic water-filling 4, 46
subchannel 19
successive interference cancellation, SIC 58